Achieving a Just Transition to a Low-Carbon Economy

Raphael J Heffron

Achieving a Just Transition to a Low-Carbon Economy

palgrave
macmillan

Raphael J Heffron
Faculty of Law
University of the West Indies, St. Augustine
Port of Spain, Trinidad & Tobago

ISBN 978-3-030-89459-7 ISBN 978-3-030-89460-3 (eBook)
https://doi.org/10.1007/978-3-030-89460-3

This Palgrave Macmillan imprint is published by the registered company Springer Nature Switzerland AG.
The registered company address is: Gewerbestrasse 11, 6330 Cham, Switzerland

Preface

I have for many years wanted to write this book and here it is finally. The end of a journey in some ways but really just the beginning in other ways. The key issue for me about the just transition to a low-carbon economy has always involved the question of "how do we achieve it?". To answer this question at first instance is to explore the legal structures needed to deliver a just transition, and that is the primary focus of this research in many ways.

My journey on the just transition began really from 2015 and seeing its mention within the Paris Agreement. I had also for some whilst thought of the wider impact of the energy sector on society and how this could be called. I was fortunate to develop ideas on this and receive funding from the UK Energy Research Centre in 2017 for developing a just transition network which still continues today. Subsequently, in 2018 I was ready to build on this, and my application for Jean Monnet Professorship in this area was accepted and awarded.

These were highlights no doubt, but together with a collective group of researchers, we held a first just transition conference in March 2018 in Edinburgh during the infamous snowstorm the Beast from the East. In March 2019 we held another event in association with the University of West Indies out in Trinidad which was also hugely popular. Our 2020 event scheduled for Edinburgh had to be cancelled, but hopefully we will return to in-person events in due course. The community of just transition scholars is now global, and researchers are working on this topic from across the world. I have over the years been involved in a range of conferences, events and invited speaker events on the just transition, and the momentum continues despite the pandemic.

In this context one of my key messages is that this challenge of a just transition to a low-carbon economy is one of the societal challenges of our time. My aim here in this text is to deliberate and discuss this collective aim and also discuss what this word "justice" means. It is vital that the transition to a low-carbon economy does not repeat the same mistakes as previous societal transitions but is one with justice at its core. Hence, I hope you enjoy this book, where I aim to explore what this justice is, who should be working on this idea, a method to utilise, legal possibilities and some financial elements for it.

St. Augustine, Trinidad & Tobago — Raphael J. Heffron

Acknowledgements

I would like to thank all who contributed in some form to the development of this short book.

I would like to thank those who introduced me fully to the energy sector at the University of Cambridge (UK) and in particular, Professor William Nuttall and Professor Angus Johnston. Thanks also to those at the various institutions where I built up my knowledge of planning, environmental, climate and energy law and policy—these include the Energy Policy Research Group (University of Cambridge, UK), Massachusetts Institute of Technology (USA), the University of Texas at Austin (US), the British Institute of International and Comparative Law (UK), the Honourable Society of King's Inns (Ireland), Trinity College Dublin (Ireland), University of St. Andrews (UK), University of Stirling (UK), the University of Leeds (UK), Queen Mary University of London (UK) and the University of Dundee (UK).

I also convey thanks to people I have worked with over the years from across the world and who I have learnt more on how the energy sector works and the justice issues that arise in the USA, Europe, Asia-Pacific, Africa, Middle East and North and Latin America.

Also, I would like to thank, in particular, the European Commission Jean Monnet Professorship award scheme where in 2019 I was successfully made a Jean Monnet Professor in the Just Transition to a Low-Carbon Economy. My thanks goes to different organisations around the world where I have been invited to present on this topic with one of the most inspirational been at the United Nations Headquarters in Switzerland in 2019 and online with them in September 2021. Thanks also goes to my students who have helped me develop my teaching in this area and have provided some new learning to me in their thoughts, arguments and assignments—this text is for you also.

A penultimate thanks goes to several research assistants John O'Boyle, Cara McKinney, Marie Tritz and Jamie Hunt who provided and completed great work in the drafting of different chapters for this book and I hope who go on to have successful careers. It is with thanks to the Erasmus programme from the EU for the funding for the research assistants.

Sincere thanks also goes to my family for their assistance and understanding and the writing of this short text in Achill, Cork, Dublin, Dundee and now in St. Augustine. And finally, this book is:

Le haghaidh mo h-oileán

Contents

1 Introduction: The Just Transition to a Low-Carbon Economy 1

2 What is the "Just Transition"? 9

3 The Just Framework 21

4 The Advance of Just Transition Commissions 43

5 Just Transitions Around the World 87

6 The Elements of the Just Transition Within International Institutions 125

7 Conclusion: The Just Transition Movement Post Pandemic, COP26 and the Financial Crisis 151

Index 159

About the Author

Raphael Heffron is dean of the Faculty of Law at the University of the West Indies, St. Augustine, and also Professor of Global Energy Law & Sustainability. He is also Jean Monnet Professor in the Just Transition to a Low-Carbon Economy awarded by the European Commission (2019–2022). In 2020, he was also appointed as Senior Counsel at Janson law firm in Brussels (Belgium). Heffron is a qualified Barrister-at-Law, and a graduate of both Oxford (MSc) and Cambridge (MPhil and PhD). His work has a principal focus on achieving a sustainable and just transition to a low-carbon economy, and combines a mix of law, policy and economics. He has over 170 publications of different types to his credit and is the most cited scholar in his field worldwide (1750+ Scopus) with translated work in multiple languages including Chinese, Italian and Persian. Heffron has given just over 160 keynote or guest lectures in 51 countries worldwide. Recent research in 2021 includes a publication in world-leading journal *Nature Energy*.

Heffron's energy research has been funded by UK national research councils (the ESRC and the EPSRC), the EU and currently through the European Commission Jean Monnet Professorship 2019–2022, UK Department for Business, Energy and Industrial Strategy (ERA-ACTOM) and the British Academy. He has given expert professional advice on energy law and policy issues (low-carbon energy and electricity systems) to the EU, UN, Commonwealth Secretariat, World Bank, and various international thinktanks—he currently serves on the UNECE Group of Experts on Cleaner Electricity Systems. More recently, he has worked on low-carbon energy projects for the 10 Member States of the ASEAN

Center for Energy; he has also produced policy reports on the energy transition and energy finance for the 54 Member States of the Commonwealth and advised Nigerian Ministries on oil and gas law and Colombian professionals on climate and green finance.

Heffron was elected to the Royal Society of Edinburgh Young Academy of Scotland in 2018. His research and teaching has been recognised by the award of a Jean Monnet Professorship by the European Commission twice in 2016 (–2019) and 2019 (–2022). His teaching has also been recognised by the UK Higher Education Academy in 2018. In addition, Raphael is a Fellow of The Royal Society of Arts.

He is a reviewer for the next IPCC Report and he is also the Consulting Editor of the current *Halsbury's Laws of England* volumes on Energy Law (the leading practitioner energy law series). Raphael is co-editor of the forthcoming *Oxford University Press Handbook for International Energy Law.* Raphael is also on the Editorial Board of the *International Energy Law Review, the Renewable Energy Law and Policy Review* and law, energy law and energy policy journals in Russia, India and Indonesia.

Raphael currently holds or has held the following positions in the past: the Co-Chair of the UK Energy Law and Policy Association; Visiting Professor in Energy Law at the International Hellenic University (Greece); an Associate Researcher at the Energy Policy Research Group at the University of Cambridge (*current*); a Visiting Professor at Paris-Dauphine University (Paris, France—*current*), Queen Mary University of London (UK—*current*), the University of West Indies (Trinidad-based), Javeriana University (Colombia), University of Western Australia (current), University Eduardo Mondlane (Mozambique—*current*) and University of Brawijaya (Indonesia—*current*); a Visiting Professor at the ESCP Business School (London and Paris—current) and a Visiting Scholar at MIT (US), University of Sydney (Australia), University Of Texas at Austin (USA) and the British Institute of International and Comparative Law (UK).

Heffron read for his PhD at Trinity Hall, University of Cambridge. He is also a trained Barrister-at-Law and was called to the Bar in July 2007 in the Republic of Ireland. He holds degrees from the University of Cambridge (MPhil-Darwin, PhD-Trinity Hall), the University of Oxford (Christ Church-MSc), the University of St. Andrews (MLitt) and Trinity College Dublin (BA, MA).

List of Figures

Fig. 1.1	Infographic on the just transition to a low-carbon economy. (Source: Created by R. Heffron 2020)	7
Fig. 2.1	Event analysis: the use of energy, environmental and climate justice. (Source: Constructed by Heffron in 2017)	14
Fig. 2.2	The legal geography "Just" Framework for the just transition. (Source: Constructed by Heffron in 2017)	16
Fig. 3.1	The Just Framework. (Source: Created by Heffron (2020))	23
Fig. 5.1	Progress towards SDG 2020 milestone. (Source (UNESCAP, 2021))	98

CHAPTER 1

Introduction: The Just Transition to a Low-Carbon Economy

Abstract This opening chapter explores the just transition to a low-carbon economy and its origin. It examines the policy journey and how it has developed since 2015 with the 2015 Paris Agreement. It highlights initial financing for the just transition and some of the key reasons why and how policy direction is needed in the area. The chapter states what will follow in the text and also introduces an infographic for the just transition to a low-carbon economy.

Keywords Just transition • Just transition journey • Just transition financing • Just transition infographic

1.1 Introduction: The Just Transition Policy and Concept

The term "just transition" has increased in popularity in academic and policy-making literature over the last few years. Its core focus has been the societal need to have a just transition to a low-carbon economy. There has been a clear and obvious divergent approach between understanding what the just transition is between the research and the professional community. For example, academia focuses on the concept, while professionals focus on what it means in terms of the changes in society.

R. J. Heffron, *Achieving a Just Transition to a Low-Carbon Economy*, https://doi.org/10.1007/978-3-030-89460-3_1

Just transition policies are still at their formative stages; nevertheless, it is important to consider their short- to medium- to long-term impact. The essence of the perspective advanced here is that there is an underlying problem to the societal transition to a low-carbon economy and that is that there is a lack of consideration of how this transition will be achieved in a "just" way. For many the idea of justice collectively and most simplistically means more fairness and the search for sustainable and equitable solutions and equality and inclusiveness. The latter focus on inclusiveness has been broadened, and now the popular phrase utilised is that the just transition is about "not leaving anyone behind".

An assessment of just transition policies to date reveals staggered plans to phase out the fossil fuel industry beginning with a coal phase-out and some countries starting an oil and gas phase-out. This is where the short- to medium- to long-term impact of such policies will potentially inhibit societies reaching the goal of a low-carbon economy. In making government interventions to support the coal industry in a phase-out, this goes clearly against the practice of competition in energy markets, particularly, in electricity markets, and remains an issue even if countries do not have competitive markets, that is, there are state-owned enterprises in operation. Nevertheless, the key issue is if there is wide-scale support to transition the coal industry. In some cases there is, for example, until 2038 in Germany, and there the question needs to be asked, "When will similar transition support schemes for oil and gas begin and then be phased out?" If coal is circa 2040 for a phase-out, will gas and oil be 2060 and/or 2080? In essence, for how long will society have to fund a transition away from fossil fuels to the development of a low-carbon economy?

In this context this book aims to address the key question of how can society achieve a just transition to a low-carbon economy. This chapter explores the journey of the "just transition" as a policy goal over the last 5 years and of how it plays a key role in how societies can achieve a societal transition to a low-carbon economy that is centred on achieving more justice. It then explores the policy interventions that see new and continued support for the fossil fuel industry. Finally, it is concluded that there needs to be greater policy realignment, transparency and accountability and policy revision in terms of how just transition policies are funded. Then the overall structure and scope of this text is stated before an infographic presented that demonstrates the basic premise of a just transition to a low-carbon economy.

1.2 The Policy Journey of the Just Transition 2015–2020

There is a range of perspectives of what the just transition is. For different communities of people, it means different things. There is a research, a practitioner (private sector), government and policy-making perspective and a justice, technological and the general public perspective. The evolution of the term "just transition" has been increasingly discussed in the academic literature in recent times.[1] However, it is its association with the development of a low-carbon economy that has driven its emergence in the last the few years.

Initially, much of the literature about the just transition is dominated by discussions around *jobs*; and it is acknowledged that has to be a key focus in order to motivate and encourage participation in the just transition. However, the issue is more than that, and how countries achieve a low-carbon economy is really the great challenge. To meet this challenge with a "just" transition adds a layer of complexity that will need a societal shift away from the long reliance on conventional energy sources, the associated transformation of the labour force to new areas and moving all other economic sectors in society away from using fossil fuels.

The just transition policy journey really began in 2015 when it was included in the 2015 Paris Agreement. It was discussed prior to this (in particular, from a jobs perspective), but this really marked its progress in policy terms. The 2015 Paris Agreement is also significant in that never before have so many countries (188) signed and ratified an international agreement so fast.[2] It has certainly achieved an impact; however, it took several years for the policy community to realise that the issue of a "just transition" was mentioned in the 2015 Paris Agreement; it stated in terms of jobs new and old.[3] It was then at the G7 talks in 2018 that momentum began to gather on the just transition where it was specifically mentioned as a policy goal in the final communique issued by the governments.[4]

In 2019, some countries began to deliver further on their just transition policies. For example, some countries have put in place the first legislative steps to achieve it. This is through the creation of what can be classified as a *Just Transition Commission (JTC)*. There are a number of different forms, but they all involve the establishment of some form of commission. Across a range of different countries, a form of JTC is in place including in Canada, Germany, Scotland, Australia, Ireland, New Zealand, the US (Appalachia), South Africa and the EU. Such commissions are established to provide expert advice on the ways to achieve a just transition and also will monitor the effects of existing laws and policies to ensure they contribute to the delivery of a just transition.

In 2020 the key change in terms of just transition policy has been the discussion and also planned allocation of funding plans for the just transition. This has in essence built on the 2019 legal developments and specifically has begun in the EU, Germany and Scotland. The actions include primarily the following, and these are significant amounts:

- EU—€57 billion has been allocated to the Just Transition Fund in September 2020[5]
- Germany—€40 billion+ allocated to support coal regions in July 2020[6]
- Scotland—£62 million fund announced in June 2020 for oil and gas[7]

1.3 Early Financing Issues in the Just Transition

The key just transition policy initiative that threatens to derail societies' plans to achieve a low-carbon policy is this latter new development in 2020 where finance is being allocated. The problem arises here in terms of the signal this financial support sends to investors. At first instance, it sends a signal to investors in coal. For the coal sector in the German example, it highlights how coal will continue in the country for a further 18 years; and there is a subsequent signal to the international community that high-income countries are still supporting coal.

There is then the key signal sent to oil and gas sectors. If coal is to be supported until 2038, will it be after that point or at what stage will oil and gas receive subsidy support? It is certain that these oil and gas sectors will seek finance for their transition—and they have been in Scotland (in the UK). Countries which develop all three of these energy sources, oil, gas and coal, will need to develop just transition financing plans, and it is very conceivable that these will stretch from 20 years up to 60 to 80 years. Given that Germany, considered a developed country, will take 18 years for its coal transition, what can be expected of some developing countries in their plans? Consider Indonesia, for example, which produces all three of these energy sources; if they are to adopt a just transition plan, it seems certain to stretch some distance into the future.

Leadership is not quite clear from developed countries on the issue of this just transition policy. Indeed, one of the hurdles in the development of renewable energy across the world was that it was not cost-competitive. The EU is a classic case of this, where subsidies given to renewable energy were withdrawn in multiple EU countries.[8] There was a worry that renewable energy companies were receiving too much subsidies; however, the development of a renewable energy was overlooked in favour of looking specifically at short-term gains by these renewable energy companies. Renewable energy was asked to compete without subsidy or with time-limited subsidies in electricity markets. And this was the case despite the demand for a low-carbon economy.

Society in essence had reached a point it where it is refusing to subsidise renewable energy which would enable the just transition to a low-carbon economy happen in a faster way. Now within this, the same timeframe society now supports just transition policy which will subsidise the fossil fuel industry. A serious question has to be asked as to why market forces are not applied to the fossil fuel industry. Should this industry not realise it is in decline in terms of the market for two reasons, which are: (1) renewable energy technology is reducing in cost; and (2) because public policy towards the product it produces is overwhelmingly negative and hence new legislation is driving up costs further and making it additionally uncompetitive.

A key reason to write this text is not only to explore the development of this transition to a low-carbon economy but also to examine how it will happen in a "just" way. To ensure the transition happens, no one can be left behind; it does concern a complete societal shift. This text approaches the just transition topic from a legal perspective—and is the first to do so—and as a result it explores how law can despite differences in thinking from different stakeholders contribute to achieving a just transition to a low-carbon economy.

1.4 Legal Redirection Needed to Ensure a "Just" Transition

One of the main challenges to achieving the just transition to a low-carbon economy is how society finances the required activities. As societies around the world face certain financing decisions in the economy (and particularly so, in light of COVID-19), there will be calls for low-cost ways of achieving policy. Unfortunately, the debate remains globally that oil, gas and coal are cheap and beneficial to the economy, and today there is a need to support them in a just transition. However, this perspective needs to change, and there needs to be more supportive action for low-carbon activities.

If society is to continue to send "supportive" signals to oil, gas and coal industries, there will be less willingness to support investment in the low-carbon transformation. A redirected effort in the allocation of investment in the just transition is needed. More research is needed on how the social transformation will happen including issues such as envisaging what society will look like, how the business-as-usual approach will change and how the law will ensure the process is inclusive (i.e. just, fair and equitable). More work is needed on presenting different pathways/scenarios of what a just transition to a low-carbon economy can look like. There are distinct trade-offs that will need to happen, and these need to enter the debate rather than be avoided.

Fossil fuels have to be allowed to suffer the fate of the market and not be protected indefinitely. If society and the international community is serious about achieving a just transition to a low-carbon economy in stages by 2030, 2040 and 2050, then action is needed now. Just transition policies around the world need to rethink their plans for subsidy support to these technologies that belong in the past.

Finally, in looking forward to a post-COVID-19 world, should society not consider a widespread massive distribution of clean energy technology and dramatically cut deaths and associated damage and costs due to climate change? Is that not where the next global societal action be?

1.5 The Structure and Scope of This Text

This text as stated earlier addresses the just transition to a low-carbon economy from a legal perspective and is the first in the world to do so. Chapter 2 covers two key issues with the first assessing what is the just transition. The second highlights the need for a convergence of communities of scholars and calls for more collaborative scholarship between climate, energy and environmental scholars. Chapter 3 builds on the previous chapter and goes further into depth on the Just Framework that can be applied in studying issues in the context of the Just Framework.

Chapter 4 then discusses one of the key ways that governments around the world are trying to achieve the just transition and that is through the use of just transition commissions—a name that can be given broadly to this type of new public administrative unit that has been created in many

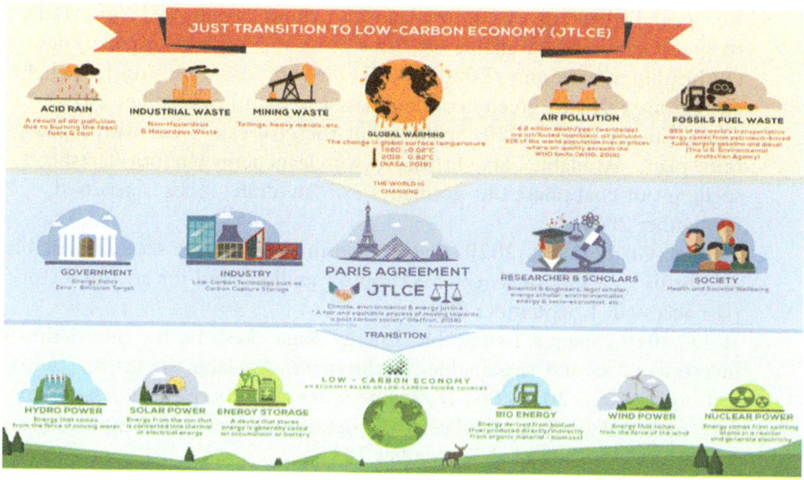

Fig. 1.1 Infographic on the just transition to a low-carbon economy. (Source: Created by R. Heffron 2020)

jurisdictions. Chapter 5 builds on the previous chapter by exploring just transition movements around the world region by region. In the penultimate Chap. 6, the focus is on international institutions and their focus on the just transition. Chapter 7 brings this short text to a conclusion with a focus on where next for the just transition taking into account the myriad of influences currently today with UN COP26 coming up, the UN-stated climate emergency and the global COVID-19 pandemic (Fig. 1.1).

Notes

1. Heffron R. J. and McCauley, D. 2018. What is the 'Just Transition'? Geoforum, 88, 74–77.
2. UN. 2020a. 2015 Paris Agreement. Available at: https://unfccc.int/process/the-paris-agreement/status-of-ratification (last accessed, 1 September 2021).
3. UN. 2020b. 2015 Paris Agreement. Available at: https://unfccc.int/files/essential_background/convention/application/pdf/english_paris_agreement.pdf (last accessed, 1 September 2021).
4. European Council. 2018. The Charlevoix G7 Summit Communique. Available at: https://www.consilium.europa.eu/en/press/press-releases/2018/06/09/the-charlevoix-g7-summit-communique/ (last accessed, 1 September 2021).

5. European Parliament. 2020. Just Transition Fund: helping EU regions adapt to green economy. Available at: https://www.europarl.europa.eu/news/en/headlines/economy/20200903STO86310/just-transition-fund-help-eu-regions-adapt-to-green-economy (last accessed, 1 September 2021).
6. Clean Energy Wire. 2020. Spelling out the coal exit – Germany's phase-out plan. Available at: https://www.cleanenergywire.org/factsheets/spelling-out-coal-phase-out-germanys-exit-law-draft (last accessed, 1 September 2021).
7. Scottish Government. 2020. £62 million fund for energy sector. Available at: https://www.gov.scot/news/gbp-62-million-fund-for-energy-sector/ (last accessed, 1 September 2021).
8. IISD. 2019. Spain's Renewable Energy Saga: Lessons for international investment law and sustainable development. Available at: https://www.iisd.org/itn/en/2019/06/27/spains-renewable-energy-saga-lessons-for-international-investment-law-and-sustainable-development-isabella-reynoso/ (last accessed, 1 September 2021).

CHAPTER 2

What is the "Just Transition"?

Abstract One of the clear areas for just transition scholarship to work on is convergence. For example, there are three research areas which all have their own version of "justice scholarship"—that is, energy justice, environmental justice and climate justice—and these now confuse and distort the transition literature. To date there has been very limited research in uniting these perspectives, and it is advanced here that a just transition can encapsulate all three perspectives, and this is where research should continue on exploring in the future. This chapter explores this issue and highlights how with a focus on developing a low-carbon economy has to unite all scholars.

Keywords Just transition concept • Energy justice • Environmental justice • Climate justice • Legal geography • Just Framework

2.1 Introduction: What Is the *Just Transition*?[1]

Within climate, energy, and environmental (CEE) scholar communities, transition means different things because each conceptualises "justice" in distinct ways. Therefore, in ensuring a "just transition" to a low-carbon economy, how can society support such a process when there are mixed visions of its meaning? These three research areas all have their own version of "justice scholarship"—that is, energy justice,[2] environmental

R. J. Heffron, *Achieving a Just Transition to a Low-Carbon Economy*, https://doi.org/10.1007/978-3-030-89460-3_2

justice[3] and climate justice.[4] At their simplest, these three forms of justice can be defined as: (1) climate justice concerns sharing the benefits and burdens of climate change from a human rights perspective; (2) energy justice refers to the application of human rights across the energy life cycle (from cradle to grave); and (3) environmental justice aims to treat all citizens equally and to involve them in the development, implementation and enforcement of environmental laws, regulations and policies. To date there has been very limited research in uniting these perspectives, and we advance a conceptualisation, the "just transition", which encapsulates all three perspectives.

In critically reviewing the transition literature, it is shown how scholars in the CEE are distorting attempts at achieving a just transition within societies. In considering space and time, there needs to be a realisation of when justice concepts within the CEE research scholarship apply. Advanced here is also the emerging area of legal geography, which allows for interdisciplinary scholarship on the concept of justice, as it applies across space and time, and in particular in relation to CEE.[5] Finally, forms of justice, that is, whether, energy, environmental or climate, need to relate to society more clearly, with the concept of the just transition offering a path forward as we hope to trace and show.

2.2 Why the Need for a United *Just Transition* Concept?

There are many reasons why there needs to be a united *just transition* concept, and a number of the most important are detailed below. In researching in the CEE area, there needs to be a realisation of realities of the world, that is, that research needs a global perspective, and actions even at a local level have national and international effects. In CEE research this issue has been identified recently[6] whereby it is acknowledged that environmental justice scholars need to ensure their research scope is global and human right-focused rather than just civil rights and locally focused; and this is similar to energy justice scholars.[7] And in a similar way, scholars have called for climate justice scholars to have a more global perspective.[8]

In considering notions of time and space, the world is witnessing an acceleration of events in different locations of the world that demonstrate that too many damaging events are occurring—such as the seven climate records broken in 2016. The just transition captures the "just" process when societies move towards an economy free of CO_2 emissions. Justice is an important element to the transition, because often the rhetoric of governments, companies, institutions and researchers discuss "a transition to low carbon economy" and then there is no mention of "just".

Transitioning away from fossil fuels in society, however, is proving to be very difficult and slow. For example, in 2016, fossil fuels accounted for 81.5% of the UK's primary energy needs, down only half a per cent from 2015.[9] Consider other examples from the UK in relation to investment in energy infrastructure and also foreign aid: in 2016, £18.6 billion (10.3% of total investment in the UK) was invested, of which 34% was in oil and gas extraction, 54% in electricity and 11% in gas, with the remaining in coal extraction and coke and refined petroleum products industries.[10] Now while the amount of investment in electricity is not presented in more detail, considering the majority of the electricity sector (54%) is fossil fuels,[11] one could make the assumption that the majority of this investment is similarly towards fossil fuels. Further, the continued support to fossil fuels by the UK is exemplified by UK foreign investment policy where through development aid, the UK supports, by a ratio of nearly two to one, fossil fuel projects.[12]

Globally, there are also problems as was mentioned earlier that identify the need for change. For example, seven climate records were broken last year in 2016: melting of Arctic ice, consecutive hottest months, hottest day in India ever, highest temperature in Alaska, consecutive and biggest annual increase in CO_2, hottest Autumn in Australia ever and highest amount of destruction in Australia's Great Barrier Reef.[13] Further, coal plant construction is on the increase, particularly, in many developing countries. Indeed, there exists far too much fossil fuels in the global energy system.[14] Hence, at an international level, it should be acknowledged that the transition needs to happen at an accelerated pace, while its slowness, however, is the norm.[15]

One of the problems for the transition is the focus of CEE research and the economic focus of the transition. Traditional economics has not really delivered positive "just" outcomes for society. If anything, it has significantly added to societal inequality; and in terms of traditional economics, it is held here that the neoclassical school of thought still dominates economic policy-making, and this echoes with research done in the fossil fuel community too.[16]

Hence, one could ask why has there not been more reform? Why has society not moved on from an economics-driven transition? Some of answers are clear, for as society was faced with powerful elites, it struggled to reform. For example, the failure to reform the financial system post-crisis of 2007–2009 demonstrates the willingness of societies to accept ongoing policy failure. Clear parallels of what society will accept is evident in the ongoing failures of policy in CEE areas, for example, recent international debate and change of rhetoric where the energy source "gas" is now classed as a "cleaner fuel" or even "transition fuel" and a lower-carbon energy source.

However, despite the powerful influence of research elites and traditional economic-driven policy-making, these are surmountable challenges. It should be recognised that these two latter groups have had notable success in how they have achieved clarity around their research, and this in turn has increased the public acceptance and understanding of their work. Unfortunately, the majority of different communities within CEE research areas all treat their concepts as separate constructs despite time and space being in a state of constant evolution and that they have a similar goal of a low-carbon economy. Indeed, it is time for more work to provide a united perspective on justice scholarship in the three areas of CEE, and it is through the just transition concept that this can be achieved.

2.3 Merging of the Three CEE Justices Under the Just Transition Concept

The areas of CEE all have their own forms of justice as stated earlier. Each is a concept widely accepted by researchers in these areas, but to date there has been little research in connecting all three forms of these justices. The just transition approach involves stakeholders of all types to the transition process. This article advocates that the just transition concept provides a more inclusive approach and as a framework encapsulates all three CEE justice communities. Further, the just transition is a concept that all stakeholders can engage with rather than having to understand all three forms of CEE justice.

There is a need to debate, discuss, research and apply the just transition. Governments worldwide are utilising the term (or words to the effect of) "transitioning to a low-carbon economy". This latter term is promoted by the *status quo*, that is, those in the dominant position in society. This is because the "low-carbon economy transition" has and will allow for a very slow transition and also one that favours this *status quo* and consequently will result in a continuation of the ongoing inequality in society.

The benefit of a just transition is its aim is to reduce inequality in modern society, which it achieves by applying justice in the areas of CEE. Inequality in society is increasing worldwide, and it represents one of the major research challenges in present-day research scholarship across many disciplines.[17] With inequality continuing to increase in society, policy reform to correct inequality clearly represents an example of policy failure. Just transition scholarship can contribute to remedying this policy failure and also contribute to scholarship on reducing inequality. An example of this is from the leading economist Thomas Piketty who as part of a research team demonstrated that there is a link between the increase of CO_2 emissions and the inequality in society.[18]

Overcoming the "inequality" issue needs all three CEE justice areas to have a clear end result. They should have a more normative and holistic view of society and how they therefore contribute to a just transition. The success of CEE forms of justice is open to question, and we need to ask, "What have they achieved to date?" Given the continued problems that society faces, it could be argued that all three forms of justice have had very limited success to date. Inequality and the ill effects of events in relation to CEE continue to happen worldwide in the developed and the developing world.

It is advanced here that a more united approach by these three CEE justice research communities could have more impact. Too often with these perspectives, there is a limited focus on the origin of the event that leads to inequality and injustice. In particular, this is evident in relation to climate and environmental justice where the focus is on adaptation, that is, after the bad "event" has occurred already, solutions are then discussed as to how to reduce the damage. Energy justice, it should be stated, for some scholars at least aims to address inequality and injustice before the "event" happens.

Overall, all CEE forms of justice need to focus more on the "event" that is under research, or the event that has triggered the research. In this study of an "event", or series of events, a greater focus on time and place is needed. For example, with time, the issues are the "pace" of change and the "timelines" of the transition. In terms of place, where these events happen and to what locations do inequalities and injustices reach or occur are important.

↓ Event Timeline ↓		
(1)Event planned & happens	(2)Damage occurs	(3)Future Impact (& which is far away)
↓ Form of Justice researched ↓		
(1)Energy Justice	(2)Environmental Justice	(3)Climate Justice
↓ Event Time & Impact ↓		
(1)Short-term	(2)Medium-Term	(3)Long-term
↓ Space (i.e. location)* ↓		
(1)Local	(2)National	(3)International
*all events will have a local, national, international element to them but in general the public will interact more with events that concern societies just transition to a low-carbon economy as this will impact on their income and wellbeing to a greater extent		
Diagram Explanation All the #1s are related as are the #2s and the #3s		

Fig. 2.1 Event analysis: the use of energy, environmental and climate justice. (Source: Constructed by Heffron in 2017)

At some different points in the analysis of the "event", the different forms of CEE justice are more relevant; however, the focus of the scholar should be on a more holistic analysis of the event and its contribution to a just transition. This process of thinking is represented in the diagram below in Fig. 2.1. For example, energy justice becomes relevant before or when an event is happening, and there is a short-term focus, that is, the aim is to change a decision as to whether something will or will not happen at a particular energy infrastructure site. This diagram is just an example, but it aims to highlight that the different CEE forms of justice have traditionally become relevant at different times over an "event", and the just transition can bring a more united and complete perspective and encapsulate all three at the same time.

2.4 Application of Legal Geography

In researching on just transition issues, there are two main academic literatures that engage with it, and these are law and geography; however, it is increasingly becoming an interdisciplinary research area. Law with "justice" and geography is where the focus is on where and when the impacts of injustices are major contributors to transition research. It is important when conducting interdisciplinary research to make connections across disciplines, and there is already a clear relationship between law and geography through the emerging field "legal geography".

Legal geography provides a useful approach for the study of the just transition. Legal geography captures research on people, space, time and law.[19] It is an area that has been researched since the 1980s and it is enjoying a resurgence and should do so in particular due to CEE issues and therefore the just transition.

The reason for thinking of the adoption of a new and more formalised way of thinking (i.e. through legal geography) is that there exists significant research in the CEE justice literature that avoids discussion on the legal context of "justice". Too often, it is not clear in this CEE justice research: (1) what justice is needed and/or expected and (2) how this will be enforced and/or applied. Law works in a system, and without the application, for example, of restorative justice, then the application of justice that enables greater distributional justice may never have the expected effect. For too long it is clear that researchers have advanced the application of different forms of justice through one of the CEE areas only not to elaborate on how and/or whether in reality it could be achieved; and a very similar criticism is noted by other scholars.[20]

This can be further evidenced by the following example. It should be recalled that the origin of environmental justice was in social protests in the US.[21] Justice was viewed to be applied if the protesting group were successful, and they forced a legal and/or policy change to a project, that is, there was a result. There needs to be a return to that clarity of thought with the just transition. More results are needed, that is, two recent results, for example, from the UK in relation to a just transition are the phasing out of coal plants by 2025 and diesel cars by 2040. These are both positive results for a just transition in the UK. In comparison, there are negative results for the just transition in the following two examples: (1) in 2017 the UK House of Commons announced that an electricity capacity auction should be technology neutral; hence, it will favour fossil fuels, which can bid in cheaper; and (2) as highlighted earlier, the UK supports fossil fuels through foreign aid.

Those previous UK examples identify a clear imbalance in overall policy towards a just transition. However, the benefit of a framework in legal geography is that it permits the researcher to identify the problems and provide research- and policy-led solutions. Figure 2.2 outlines below a legal geography perspective on the just transition. The framework advanced is quite simple, the "Just' Framework. It borrows from law and geography

J	TRANSITION	Justice	Justice takes the form of 3 forms of justice: Distributional; Procedural; Restorative
U		Universal	Universal takes the form of two universal forms of justice: Recognition; Cosmopolitanism
S		Space	Space brings in location, where are 'events' happening ? (in principle, at local, national and international levels)
T		Time	Time brings into transition timelines such 2030, 2050, 2080 etc. and also 'speed' of the energy transition (i.e. is it happening fast enough?).

Fig. 2.2 The legal geography "Just" Framework for the just transition. (Source: Constructed by Heffron in 2017)

and also the three forms of justice across CEE. It is outlined below, and many will be familiar with the types of justice mentioned below, which are explained already in the literature. Future research will identify them in more detail later as they apply specifically to the just transition.

2.5 Steps in the Future Development of the Just Transition

In thinking of the evolution of the just transition and also the justices of CEE, it is clear that there needs to be a measurable outcome from an action. The measure in the majority of these forms of justice is not quantitatively measured but can be assessed through whether it results in a direct change in a law and/or policy or, more simply, where a previous decision is overturned or improved. Over time scholars have worked on justice in their research, but there has been limited scholarship as to the broader goal of a just transition. A just transition is a societal goal, and the three communities of CEE justice scholars need to think together and have as their common purpose a just transition.

To ensure the debate around a just transition and its effect on policy-making are not another policy distorted by the traditional economic thinking, there needs to be a more united framework across CEE justice. To move society towards a just transition, the importance of public acceptance and understanding is crucial to ensure they act towards and support such a transition. And a just transition concept that unites the CEE justice fields can result in significant increases in public acceptance and understanding.

Time is of the essence for society, and an accelerated just transition is needed, given all the issues in society related to CEE sectors. The question has to be asked, "What have the different forms of CEE justice achieved?" Albeit energy justice is a newer concept, it is also strange that it is so, since the energy sector is primarily responsible for the CO_2 emissions which is the main focus of environmental and climate justice scholarship. Further, many sectors in the economy rely heavily on energy, such as transport, industry and even human effort (energy from food provided by the energy-intensive agricultural sector). The application of justice in the energy sector has clearly been a forgotten issue, and the dominance of economics and big energy companies may explain why it was not a feature until recently; for a recent account of the development of energy justice, see the *Concept of Energy Justice* across the discipline's paper.[22] The just transition can account for these issues and encapsulate those from across CEE justice scholarship.

It is advanced here that legal geography, which incorporates two key disciplines working together on a framework on the just transition, can deliver more realistic research conclusions, and therefore more achievable law and policy options. Research on a just transition needs to learn from the limitations of CEE justice research and refocus on being result-driven since the speed of the transition is a major concern globally.[23]

The major contribution of uniting CEE justice is the added potential to contribute to reducing inequality and injustice in society as it relates to CEE sectors and aims to achieve a just societal transition. Both law and geography have long histories in research on inequality and injustice; it is time to ensure they are added to the debate on CEE issues and that the dominance of traditional economic thinking is reduced. The *status quo* in relation to issues in CEE policy and sectors needs to be challenged in research and practice. Researchers on CEE justice need to integrate their work into the bigger picture, that is, the just transition, and they must focus on the implications of their work for this just transition to low-carbon economies.

Notes

1. A variation of this chapter originally appeared in 2017 with a publication date of 2018 as a short critical review in the journal *Geoforum*. It is with thanks to them and my co-author for the article that it is here. Original

reference and a variation is here: Heffron R. J. and McCauley, D. 2018. What is the 'Just Transition'? Geoforum, 88, 74–77.

2. McCauley, D., Heffron, R. J. Stephan, H. and Jenkins, K. 2013. Advancing Energy Justice: The triumvirate of tenets. *International Energy Law Review*, 32 (3), 107–110.
3. Capek, S. 1993. *The "Environmental Justice" Frame: A Conceptual Discussion and an Application.* Social Problems, 40 (1), 5–24; and Walker, G. and Buckeley, H. 2006. *Geographies of Environmental Justice.* Geoforum, 37 (5), 655–659.
4. Caney, S. 2014. *Two Kinds of Climate Justice: Avoiding Harm and Sharing Burdens.* The Journal of Political Philosophy. 22 (2), 125–149.
5. For example, see: Blacksell, M., Watkins, C. and Economides, K. 1986. *Human geography and law: a case of separate development in social science.* Progress in Human Geography, 10 (3), 371–396; Delaney, D. 2003. *Beyond the word: law as a thing of this world.* In: Holder, J. and Harrison, C. E. (eds) *Law and geography.* Oxford, UK: OUP (pp. 67–84); and Blomley, N. K. 1994. *Law, space, and the geographies of power.* New York, US: Guilford Press.
6. Agyeman, J. 2014. Global environmental justice or Le droit au monde? Geoforum, 54 (July), 236–238.
7. Heffron, R. J. & McCauley, D. 2017. *The concept of energy justice across the disciplines.* Energy Policy, 105, 658–667.
8. Forsyth, T. 2014. *Climate Justice is not just Ice.* Geoforum, 54 (July), 230–232.
9. Carbon Brief. 2017. *Six charts show UK's progress on low-carbon energy slowing down.* (31 July 2017). Available at: https://www.carbonbrief.org/six-charts-show-uk-progress-on-low-carbon-energy-slowing-down (last accessed 1 September 2021).
10. Department of Business, Energy & Industrial Strategy (BIES)/United Kingdom Statistics Authority (UKSA). 2017. *UK Energy in Brief 2017.* Available at: https://www.gov.uk/government/uploads/system/uploads/attachment_data/file/631146/UK_Energy_in_Brief_2017.pdf (last accessed 1 September 2021).
11. Department of Business, Energy & Industrial Strategy (BIES)/United Kingdom Statistics Authority (UKSA). 2017. *UK Energy in Brief 2017.* Available at: https://www.gov.uk/government/uploads/system/uploads/attachment_data/file/631146/UK_Energy_in_Brief_2017.pdf (last accessed 1 September 2021).
12. CAFOD. 2017. *UK Support for Energy in Developing Countries.* Available at: https://cafod.org.uk/content/download/27353/269740/version/2/file/Policy%20briefing%20UK%20Support%20for%20Energy%20

in%20Developing%20Countries%20Oct%202015.pdf (last accessed 1 September 2021).
13. The Guardian, 2016. *Seven climate records set so far in 2016.* (17 June 2016 – Adam Vaughan). Available at: https://www.theguardian.com/environment/2016/jun/17/seven-climate-records-set-so-far-in-2016 (last accessed 1 September 2021). This is just a newspaper report connecting to the issue – however, there are many international reports.
14. Figueres, C. et al. 2017. *Three Years to Safeguard our Climate.* Nature. 546 (7660) 593–595. (29 June 2017).
15. Figueres, C. et al. 2017. *Three Years to Safeguard our Climate.* Nature. 546 (7660) 593–595. (29 June 2017).
16. Rist, G. 2016. (4th ed.). (Translated by Camalier, P). *The History of Development: From Western Origins to Global Faith.* Zed Books: London, UK.
17. See the work of the following leading economists in this area: (1) Stiglitz, J. E. 2012. *The Price of Inequality.* Penguin Books: London, UK; (2) Atkinson, A. B. 2015. *Inequality: What can be done?* MA, US: Harvard University Press; (3) Piketty, T. 2015. *The Economics of Inequality* (translated by Goldhammer, A.). Belknap Press of Harvard University Press: MA, US; and (4) Scheidel, W. 2017. *The Great Leveler.* Princeton University Press: NJ, US.
18. Chancel, L. and Piketty, T. 2015. *Carbon and inequality: from Kyoto to Paris.* Paris School of Economics (November 2015).
19. (1) Blacksell, M., Watkins, C. and Economides, K. 1986. *Human geography and law: a case of separate development in social science.* Progress in Human Geography, 10 (3), 371–396; (2) Blomley, N. K. 1994. *Law, space, and the geographies of power.* New York, US: Guilford Press; and (3) Delaney, D. 2003. *Beyond the word: law as a thing of this world.* In: Holder, J. and Harrison, C. E. (eds) *Law and geography.* Oxford, UK: OUP (pp. 67–84).
20. Jamal, T. and Hales, R. 2016. *Performative justice: New directions in environmental and social justice.* Geoforum, 76 (November), 176–180.
21. Agyeman, J. 2014. Global environmental justice or Le droit au monde? Geoforum, 54 (July), 236–238.
22. Heffron, R. J. & McCauley, D. 2017. The concept of energy justice across the disciplines. Energy Policy, 105, 658–667.
23. Figueres, C. et al. 2017. Three Years to Safeguard our Climate. Nature. 546 (7660) 593–595. (29 June 2017).

CHAPTER 3

The Just Framework

Abstract The just transition is a more inclusive approach to the development of a low-carbon economy. It involves stakeholders of all types to the transition process. There is a need to debate, discuss, research and apply the just transition into practice. This chapter delves into depth and introduces the Just Framework. It highlights what justice is being sought in the just transition, that is, distributive, cosmopolitanism, recognition, restorative and procedural. It explores the just transition also in terms of space and time. Further, it tests the Just Framework with a small case study of critical minerals which are vital in terms of technology for the overall transition to a low-carbon economy.

Keywords Just transition • Just Framework • Critical minerals • Distributive justice • Procedural justice • Restorative justice • Recognition justice • Cosmopolitanism justice

3.1 Introduction: Background to the Just Framework

Research that is focused on the role of justice should utilise some element of legal research methodology. As part of legal methodology, there is an opportunity to engage in interdisciplinary research and which the research engages in itself—albeit many legal researchers have been slow to engage

R. J. Heffron, *Achieving a Just Transition to a Low-Carbon Economy*, https://doi.org/10.1007/978-3-030-89460-3_3

with interdisciplinary research. There are three core legal areas of research that can be highlighted, and they all feature to a degree in the Just Framework, and it is important for interdisciplinary scholars to understand the origins of this framework.

Firstly, legal research explores from a conceptual doctrinal perspective (i.e. the study of the written law itself), and from comparative legal perspective, that is, the core justice aspects of an issue are explored. Second it examines the issue from an interdisciplinary perspective (sociolegal) utilising legal, economic and business taxation perspectives to deliver on what the key challenges for achieving justice are. In essence the research here looks at the "rules of the game" created by the legal structures. Third, the underlying focus here is to ensure that the transition to a low-carbon economy happens in a "just" way (as outlined in Chap. 2), and hence the "Just Framework" will be utilised which is an interdisciplinary method (legal geography).

Therefore, for that to be achieved, all the elements of justice need to be addressed such as distributive, procedural, recognition, restorative and cosmopolitanism. This research builds on the basic Just Framework stated in Chap. 2 and develops it further and supplies an infographic to add to the explanation of it.[1]

3.2 The Just Theoretical Framework

The theoretical framework utilised here is the "Just Framework". This framework is demonstrated below in Fig. 3.1 which has been advanced previously by scholars in the area who state it combines an interdisciplinary legal geography perspective.[2] More recently, it has been used directly to research justice and critical minerals.[3] The aim of the Just Framework is to unite climate, environmental and energy (CEE) justice scholarship and as a result try to reduce inequality and injustice within society. The benefit of utilising the *Just Framework* is to ensure a practical research perspective to a problem. In this context, for example, there is a focus on considering a number of issues such as the impact of the business supply chains that critical minerals are involved in and then energy, climate and transition timelines, such as 2030, 2040, 2050 and beyond. The *Just Framework* covers five key elements of justice that are needed to ensure the application of human rights across the energy life cycle,[4] and these include briefly:

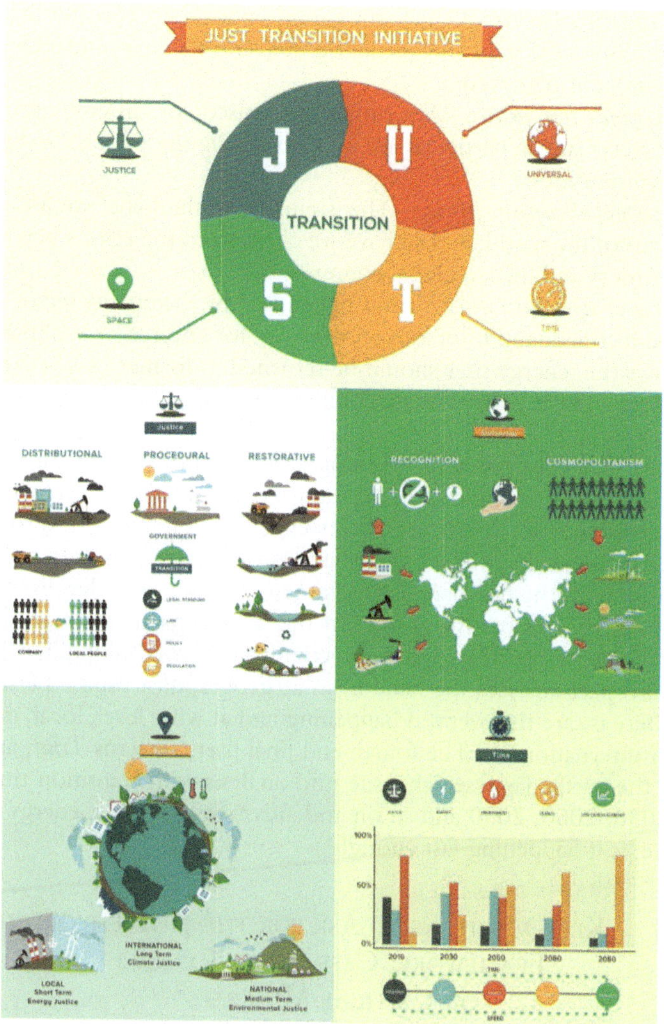

Fig. 3.1 The Just Framework. (Source: Created by Heffron (2020))

- *Distributive justice*—This concerns the distribution of benefits from the energy sector and also the negatives (i.e. are oil and gas revenues shared sufficiently? Who suffers the environmental damage?).

- *Procedural justice*—The focus here is on legal process and the necessary full legal steps (i.e. are all the steps for an environmental impact statement observed?).
- *Recognition justice*—Are rights recognised for different groups in society? (i.e. in particular are we recognising the rights of indigenous communities?)
- *Cosmopolitanism justice*—This stems from the belief we are all citizens of the world, and so have we considered the effects beyond our borders and from a global context?
- *Restorative justice*—Any injustice caused by the energy sector should be rectified, and it focuses on the need for enforcement of particular laws (i.e. energy sites should be returned to former use; hence waste management policy and decommissioning should be properly done).

Figure 3.1 explains with an infographic created for this article that explains the Just Framework. This framework has four key elements: justice, universal, space and time. The essence of the framework is to enable the researcher to identify problems and provide research- and policy-led conclusions. In the first part of the framework, they consider distributive, procedural and restorative justice. In the second part, two more holistic forms of justice are employed in recognition and cosmopolitan justice. The third part analyses the issue from a (geographical) space perspective (i.e. where is/are the event(s) happening and at what level, local, national and/or international?). The fourth and final part concerns *Time*, and this brings the particular research issue into analysis with transition timelines such 2030, 2050, 2080 and so on and also "speed" of the energy transition (i.e. is it happening fast enough?).

3.3 Key Questions to Ask for the Just Theoretical Framework: Case Study Example on Critical Minerals[5]

3.3.1 Background to the Case Study

It is important to remember that the critical minerals sector is part of the mining sector which the UN has stated clearly is crucial to meet the UN Sustainable Development Goals[6]—the mining sector includes mineral exploration, extraction, milling/concentrating, smelting, refining, processing, transport and sale. Nevertheless there is little literature on either research or practitioner that focuses on issues of justice in critical minerals development and extraction. Predominantly, practitioner and policy literature focuses in essence on (1) developing the critical minerals business, (2) rising demand, (3) raising finance and (4) political issues.[7] There is one policy report however from the EU that does highlight "fairness"; however it in effect is stating that the EU should have access to these critical raw materials at fair prices.[8] This is a very limited form of justice that seems to favour big business and industry in general.

There is other literature that has aimed to review some considerations of justice, but actual references to justice within the text are rare,[9] while other researchers have assessed justice issues in the context of a single metal such as cobalt and a focus on a single justice issue, for example, distributive justice.[10] Further, while a recent UN (2019) report does focus in a limited way on the role of law, it is mainly in the context of procedural justice and is just a general focus on the mining sector.[11] That UN report was preceded by another general mining report in 2018 that focused on access to justice which again touches on procedural and to a certain degree recognition justice.[12] In contrast, this article aims to be far more comprehensive in its focus on justice recognising the development of critical minerals is a global issue, they are part of global business value and supply chains and they play a significant role in ensuring there is a just transition to a low-carbon economy across the world.

There has however been a growth in literature on critical minerals over the last few years, and this will increase. But to date, this increase in the literature has not assessed issues of justice for this sector. The justice that needs research is for critical minerals development, and this needs to reflect their role in the transition to a low-carbon economy. As stated by Ali et al. (2017: 367)[13] in *Nature* in 2017, "a transition to a low carbon society, [is] a change that will require vast amounts of metals and minerals. Mineral resourcing and climate change are inextricably linked, not only because mining requires a large amount of energy, but also because 'the world cannot tackle climate change without adequate supply of raw materials to manufacture clean technologies'". There are other past and more recent key literature on the topic in key journals of note such as *Nature Geoscience*,[14] *Science*[15] and *Renewable Energy Strategy Reviews*,[16] but again none of this focuses directly on justice issues. Environmental and social issues are identified but limited in coverage, and this is where this article aims to develop the literature and identify how issues of justice (or fairness, equity, etc.) will be examined and resolved.

Why Is Justice Needed in Critical Minerals Development?

It seems strange to ask such a question of why justice is needed. However, it is clear that in the context of the development of critical minerals, that question needs to be asked. An example of why, is the significantly documented issue of child labour surrounding cobalt extraction in the Democratic Republic of Congo.[17] One of the reasons that law developed in terms of the extractive industries was on repeated issues of safety failure back in the early 1800s when coal began to be mined commercially across the industrial world.[18] Yet nearly 200 years later, the same issue remains in different parts of the world. Justice in reality has only touched some parts of the mining sector.

A key reason for justice is to ensure fairness and equality in society, in essence to resolve inequalities. Economics since the 2007–2009 financial crisis has started to engage on this issue in a significant way.[19] Indeed Piketty (2020: 670)[20] decries the issue and states that despite living in a world of big data, public data on inequality is inadequate. Indeed, Piketty[21] notes the key issue of inequality and climate change emissions, and from this perspective, he highlights how this will cost economies (and more likely developing ones) 5 to 20% of global GDP if not more (he cites the Stern Review (2007), and the IPCC (2018)) report demonstrates these effects may be accelerated as a result of pollution and environmental damage since 2007.

Considering the broad role that critical minerals play in the development of the low-carbon economy that the majority of countries have signed up to achieve via the Paris COP21 Agreement, it is necessary to have a broad view of justice that can encapsulate this global industry that has global impacts.

In this section utilising the Just Framework, there are key theoretical assumptions which can be advanced for ensuring and measuring the level of justice for critical minerals development. Each of the four core parts of the Just framework are examined—justice, universal, space and time—and analysed in turn. The focus here concerns that key issues that society needs to ensure are happening as the critical minerals sector develops across the world. In some cases, there are already actions happening which the critical minerals industry can adopt. The aim is not to explore every single type of issue but rather focus on one and/or several key issues under each of the four parts of the Just Framework. Finally, the contribution here in terms of presenting this case study lies in how to move from theory to practice in terms of achieving justice as the critical minerals industry develops.

3.3.2 *Justice and Its Elements: Distributive, Procedural and Restorative Justice*

Distributive Justice

Distributive justice is one of the three cornerstones of justice. There is a need to ensure there is a fair distribution of resources in society. As stated earlier, many of the leading economists have been working directly on this issue where there is not a fair distribution of resources in society and there are then resulting effects of inequality. This issue is crucial as the critical minerals industry develops. There is a need to ensure that the same issues that have arisen in the extractives industry do not arise again.

The core issue for distributive justice is taxation. There is a need to know where the revenue from the critical minerals is going and how is it distributed. Recent scholarship has concluded that distributive justice should be recognised as the cornerstone of tax law, that is, as the "first or sovereign virtue of a society's tax system" (Duff, 2017: 167).[22] It has even been advanced that Adam Smith has argued that distributive justice was a key tax policy goal and that legal certainty was needed in order for distributive justice not to be just an illusory goal and that this still applies to this day.[23] And this is certainly an issue for many developing countries when establishing their taxation policy around critical minerals.

The reality however of revenue distribution from a resource that is extracted is difficult to calculate because it is a complex operation and then there is a global market for the price of the resource; hence and too often the revenue sharing arrangements for a project are completed in the majority of projects in a case-by-case basis, despite the existence of taxation legislation. There are, however, actions on this issue, through the initiative the Extractive Industries Transparency Initiative[24] which requires a state and an investor to disclose the details of their taxation relationship. It is having an impact though there remain significant problems, with only 53 countries having signed up and not all cooperating to the same extent.

There is further global action also from an OECD-led initiative which seeks to reform issues around taxation of energy and resource extraction.[25] In addition, this OECD initiative representing circa 60 countries noted that energy and mineral resource extraction represents a significant opportunity to raise revenue for many governments. Post COVID-19 the need for more revenue will be even more pronounced, and hence it is vital that tax reform occurs at international and national levels. More transparency in taxation issues is needed at the international level, while at the national, tax reforms are needed to increase revenue.

Procedural Justice

Procedural justice focuses specifically on the legal process of taking a project through from start to finish, i.e., from planning and construction to operation and to end use, and do all stakeholders have a legitimate opportunity to be involved (or represented) at these different stages. There are of course many issues here in terms of finance and taxation which are mentioned in the elements of justice here. The key issue for the development of critical minerals industry in procedural justice is the Environmental Impact Assessment (EIA) process each country has at a national level.

This is the legal process a project developer has to go through in three phases and at an international, national and local level:

1. *International*: In securing finance for the project, this will require an EIA to be produced for the financing institution under international banking standards—the Equator Principles (this may be different for a company who will finance the project in-house or a national company who avails of financing options within the country, but the likelihood is there would remain some environmental impact statement produced).
2. *National*: Has to adhere to national EIA legislation and submit an Environmental Impact Statement (EIS) which has to be approved before the project receives permission to start.
3. *Local*: The EIA process has to include several elements of public participation and involve these local stakeholders in the plans for the development of the project.

There have been two recent cases where energy projects, coal, have failed to pass the EIA process, and these were in Australia and Kenya. A key reason for the failure in both countries was that the EIAs lacked completeness in terms of data provision, the assessment of the social and environmental impacts from the existing data was poor, and the projects' positive economic contributions were overestimated.[26]

Of significance is that the legislation around EIAs is changing periodically, and this is for two main reasons; the first is that once it changes, that is, is improved in one country, other countries follow. The reason why this happens is that that increased requirements in the EIA process reduce the risk profile of the project for all stakeholders. The second reason for change is that data capture methods are increasing so different types of environmental impact as well as socio-economic impact of a project can now be measured. EIAs are becoming more and more stringent, and this is documented in the literature; it will be increasingly hard for energy projects, including critical minerals, to pass this process.

There is additional support for EIA legislation through international law through an international agreement: the Aarhus Convention 1998—known as the Convention on Access to Information, Public Participation in Decision-Making and Access to Justice in Environmental Matters. This agreement allows for public participation in energy projects and access to environmental data and is an agreement signed by 39 countries but with 47 parties to it. The result is that there is added enforcement to national EIA legislation, but this was a convention advanced by the UN ECE so is limited mostly to EU and neighbouring countries. However, perhaps a more significant development has been in international finance standards, known as, the Equator Principles. These Equator Principles will play a significant role in the critical minerals industry as they cover the majority of international project finance debt within developed and emerging markets, and notably they in essence require the project to have passed the EIA before releasing the project finance.[27]

Restorative Justice

Restorative justice concerns how if there is an injustice in the energy sector it should be rectified.[28] This could be in the form of revenue distribution from the project (but that is covered principally under distributive justice). The key issue for the critical minerals industry would be that these energy sites should be returned to their former use. Hence waste management policy and decommissioning should be properly completed and costed within a project and the rules set out in legislation. In addition, restorative justice can aid in pinpointing where prevention needs to occur.

Of importance across the world as critical minerals are developed is that these sites of extraction are not abandoned. The energy sector already has a well-documented history of abandoned energy sites which then become environmental problems.[29] In ensuring that this does not happen in the development of the critical minerals industry and that justice has a role, a "restorative cost" needs to be built into the projects costs and that needs to be factored in during the three phases of the relevant critical mining project as outlined below:

1. *Project Development and Construction*

The EIA process needs to include the "restorative cost" and plans in the EIS that is prepared and which involves multiple stakeholder consultations.

2. *Project Operation*

As was previously the case, just satisfying the local community at the outset of the project is not sufficient, and the project operator needs to work with the local community for the duration of the project operation. This has been formalised into a Social License to Operate (SLO) agreement and in some countries is recognised in legislation.[30] Again part of the cost here would be restorative as gradually the project operations reduce over time, and these should be factored into project costs.

3. *Project Closure and End-of-Life Process*

This is a clear restorative cost and is in essence the decommissioning or waste management operational cost. This is a major area of weakness and needs to be rectified in national and international law and policy.[31]

One tool that is already in use in some countries are forms of an Energy Financial Reserve Obligation (EFRO) which can be referred to as a clean-up obligations and/or environmental bonds. In the US, under the federal Surface Mining Control and Reclamation Act (SMCRA) 1997 (and in Australia it's the Financial Assurance under the Environmental Protection Act 1994), energy companies are required to remediate the lands where mining activity has occurred. However, many companies were allowed to self-bond, and therefore when they went bankrupt, there was still no finance available for meeting reclamation obligations,[32] and the EFRO counters this corporate behaviour.

The EFRO now has become legislation in more than several countries, and it is already causing an impact on the value of an energy asset as it requires an operator to place an estimated amount of money that will pay for the decommissioning of the energy infrastructure and site into a neutral bank account; this is to avoid the scenario where the state has to pay should the operator go bankrupt or sell on the asset to a company without the financial capacity to pay for a clean-up.

3.3.3 Universal

Recognition Justice

Recognition justice is about whether rights of different groups are recognised as development happens. In particular, recognition justice in the energy sector has focused on the effect of energy developments and activities on indigenous communities.[33] There is international law that supports the rights of indigenous communities in the United Nations Declaration of the Rights of Indigenous Peoples (UNDRIP) 2007 where a key part of that treaty is Article 10—"Indigenous peoples shall not be forcibly removed from their lands or territory. No relocation shall take place without the free, prior and informed consent of the indigenous people, and after agreement on just and fair compensation and, where possible, with the option of return" (UNDRIP: 11).[34] Indeed, it is considered international law due to the reliance on it and the American Declaration of the Rights and Duties of Man (1948) by international and state tribunals.[35]

These issues arise repeatedly as a result of energy projects, and most recently, one legal case concerned directly this issue where the foreign company abused its position, polluted sacred places and committed other serious crimes—*South American Silver Limited (Bermuda) versus the Plurinational State of Bolivia, PCA Case No. 2013-15*. The final result was in 2018 where the project had to be expropriated from the investor with only sunk costs paid which was $18.7 million instead of the $385 million claimed by the investor—the case was heard under the rules of the United Nations Commission on International Trade Law. Indeed, the UN have long since recognised this issue and had a report conducted in 2012 on *An Analysis on the Duty of the State to Protect Indigenous Peoples Affected by Transnational Corporations and Other Business Enterprises* which called for significant action by governments to protect the rights of indigenous communities from resource extraction companies.[36]

In recognising the rights of indigenous communities, it is important that new legal structures strengthen their voice, and in that context it is welcome that legal agreements around a *Social License to Operate* (SLO) are on the rise. An SLO can be defined as a legal instrument for achieving justice and ensuring long-term sustainability in energy investments and a just transition to a low-carbon economy.[37] It is an issue rising to prominence, and a case finally concluded in 2017 recognised the SLO has a key challenge, and it was in part why there was the expropriation of a silver mine in Peru because the investor and local population had significant mistrust of the investor; consequently only sunk costs were paid which was $18.2 million instead of the $522.2 million claimed by investor—Bear Creek versus Peru Bear Creek Mining Corporation versus Republic of Peru, ICSID Case No. ARB/14/2. It is clear from the judgement in the aforementioned case that the SLO had been broken by the investor and project operator, and that led to the necessary expropriation of the asset.

In addition, in there is a broadening of the term recognition justice, as many advocate it is not only indigenous communities that deserve recognition for their rights but many disadvantaged communities do, as in essence this is where the issue of societal inequality rises. For example, Heffron et al. (2018a)[38] analyse mining projects in Colombia—which were critical minerals projects—and highlight how the communities worst affected are communities who are in "extreme" poverty according to the UN standards. The question that needs to be asked is whether there is enough action to ensure these local mining communities share in the benefits of these critical minerals: Are their socio-economic rights being respected?

The issue of human rights and business is a fast-growing area of research and practice in law[39] and particularly so in terms of the distribution of tax revenue and its relationship to human rights. This study by the IBA followed a previous study which examined in detail the effects of tax abuse and had a major finding that tax abuses have a significant impact on the enjoyment of human rights.[40] In addition, the recent *UN Committee on Economic and Social and Cultural Rights* report highlights the role of states and also businesses in determining the realisation of economic, social and cultural rights.[41]

Cosmopolitan Justice

In the context of critical minerals development, the link to cosmopolitan justice where it is based on the core premise that we are all citizens of the world is clear. There is a global effect to our actions, and as identified earlier, the critical minerals industry is a global industry, and these minerals are utilised in products all in use on a daily basis. There is a need to recognise therefore the multiple cross-border effects of our activities as a result of global supply and business value chains.

There is a growing recognition of this issue, and as will be outlined below, this has to be realised and accounted for as the critical minerals extractive industry develops. There are several recent clear examples where there is a growing interest in taking legal action as a result of cross-border or overseas effects. A first clear example, though not related to critical minerals, is in coal where in 2019 in Australia a judge stated in a decision that a coal mine should not receive permission to open due to the effects of the carbon dioxide that would be produced in other places in the world—*Gloucester Resources Limited v Minister for Planning [2019] NSWLEC 7.*

Indeed, disputes are on the rise in the mining sector, and increasingly it concerns some issue that is international in nature.[42] A second example that does reflect the parts of the critical minerals extractive industry specifically is the actions by the Canadian Government, where they have created and appointed in 2019 an Ombudsperson for Responsible Enterprise that will assess and investigate the actions of Canadian overseas companies focusing on human rights abuses in mining, oil and gas and garments.[43] Independent of that is that there are already three cases being taken by foreign entities against Canadian companies in Canada itself, for human rights abuses in their foreign sites in Eritrea and two in Guatemala.[44]

There are other examples of cross-border actions which can affect the critical minerals industry, and these relate to the aforementioned EIAs, Equator Principles and tax initiatives mentioned under the *justice*, first part of the Just Framework. And in addition, the 2015 Paris Agreement has already been noted, and it is worth recalling that 189 countries have signed and ratified it (197 have signed overall).[45] That is a short space of time for so many countries to sign up but emphasises the international cooperation in the world on achieving low-carbon economies. This international cooperation is also evident in another convention, the Convention on Environmental Impact Assessment in a Transboundary Context, known as the Espoo (EIA) Convention (1991). In brief, this convention concerns the obligation on states to report to each other the effects of activities within their boundaries that may have cross-border impacts. Although finalised in 1991, it was only in 1997 six years later before there were enough countries to enter it into force, and there remain only 45 signatories to it.[46] The majority of signatories of the Espoo Convention are European because it was the UN Economic Commission for Europe that advanced it. Nevertheless, it shows the success of the Paris Agreement in terms of the number of countries signing up and in such a short space of time. If the Espoo Convention were more international, there perhaps would be a rise in issues that this convention could resolve, but it is an area that threatens to grow over the coming decades, and the precedence of this convention will grow in importance. A final area of further research will concern access to critical minerals and how this can be just and fair given their location and ownership—early research has been completed on this in the form of advancing a Fair Trade mineral supply.[47]

3.3.4 *Space*

The third part of the Just Framework is "space" which covers a range of issues listed below which follow key literature in geography around space:[48]

- The location, that is, where the "events" are happening
- Whether the action is happening at a local, national and international level
- The interlinks, that is, is it part of a global supply chain, and what are the characteristics of this
- The stakeholders involved and their activities, that is, a multinational company utilising tax services, for example

It is of vital importance that the critical minerals industry is considered as a global industry. Actions that happen at a local level in terms of mining operations are all influenced by what happens at a national and international level. International taxation practice, for example, guides the taxation of the critical mineral being extracted which in turn influences the scale, length and operational design of the extraction project. In essence actions in one part of the world influence another.

This can be demonstrated in one study which looks at the issue of a cobalt business value and supply chain where the cobalt is used for battery technology.[49] The figure identifies three phases where the cobalt is extracted (1), then exported to Asia where it is utilised in battery production (2) and then eventually consumed by Western and Asian markets (3). In this context the critical minerals industry has multiple locations of "activity".

As a result of this, there are multiple issues that get raised at local, national and international levels. One of these is to explore the flow of finance around that global supply chain. The Amnesty Report (2017) highlighted the issue of labour, and in essence they state there are human rights abuses occurring in the extraction phase so hence it could be determined that the DRC is not receiving its due share of revenue from cobalt.[50] Research identifies the clear injustice in terms of the tax rate which was 6% (between 2011 and 2014) when the recommendation by the World Bank for such minerals was 46%.[51]

Further, it is important to consider then where the profits from the mining sector and as a result where profits from the critical minerals sector will end up. This will generally be in bank accounts offshore that belong to a collection of technology companies and mining companies via the services of professional firms who assist in this process. Indeed, the IGF-OECD initiative representing circa 60 countries released three reports on the aggressive tax avoidance by multinational firms in the mining sector.[52] There is no doubt that the international community is thinking about how the critical minerals industry may evolve here, and that confirms why it is so necessary to ensure justice within the development of the critical minerals industry.

3.3.5 Time

Time brings into transition timelines such 2030, 2040, 2050, 2080 and so on and also "speed" of the energy transition (i.e. is it happening fast enough?) which many say it is not[53] and hence the need for a focus on an area such as critical minerals which has a crucial role to play.[54] Transition timelines and/or energy and climate timelines have become a feature of the energy system and for the development of a low-carbon economy. Therefore, they have an immediate and significant effect on the critical minerals industry as they indicate which minerals will be in demand for the transition. A key influence on future timelines has been the success of the Paris COP21 climate change negotiations and subsequent agreement, the Paris Agreement (2015).

The Paris Agreement, while it does not specifically bind a country to meeting a goal in 2030, it does require them to produce a clear pathway of how it will achieve these goals. Some countries have been taking progressive action in meeting 2030 goals—such as the UK phasing out the use of coal by 2025,[55] but many others have 2030 goals than align with 2040 and 2050 goals, and the Paris Agreement is the first stage of this transition to 2050 (with an estimated 77 countries aiming for net zero carbon emissions by 2050).[56]

In thinking of how to achieve, for example, 2030 climate and energy goals, a country has to establish new law on how to achieve this energy transition. There is no common solution as countries have control over their own energy resources (a right of a nation in international law[57]) and they have different geographies, culture and socio-economic characteristics. However, there is a common demand over the critical minerals needed for this transition to a low-carbon economy.

Timelines are of vital importance to the transition to a low-carbon economy and in particular from a legal perspective. Law needs to be in place that can deliver on climate and energy goals for 2030, 2040 and 2050. In different countries, the formulation of law can take different lengths of time, for example, from one to five years, and that is not even to question whether that law is effective. The essence of the issue here is that in order to achieve a 2030 energy or policy goal that will deliver a transition to a low-carbon economy, the law needs to be changed, reformed and so on at least a reasonable length of time before that—one would need to think before 2022 at the latest to achieve 2030 goals. A key reason why this is necessary is to ensure that there is a stable legal structure around the activity, in this case critical minerals development. This legal stability (or what can be referred to as *legal certainty*) provides confidence to investors who can then make an investment decision knowing that they will be able to secure a return on their investment and that their investment is protected (to some degree).

Therefore, for critical minerals to be developed in a just way, there is a need for legislative development that (1) covers all issues around extraction, production and export and (2) ensures that the latter legislative development complies with and adapts to (such as adapting to the demand for critical minerals as a result of these international agreements) goals of the Paris Agreement at national levels and international energy and climate targets.

Notes

1. The work builds on the Just Framework from these papers where it is applied in the case of critical minerals which will be much needed for a just transition to a low-carbon economy: (1) Heffron, R. J. 2020. The Role of Justice in Developing Critical Minerals. *The Extractives Industry and Society*, 7 (3), 855–863; and (2) Qurbani, I. D., Heffron, R. J. and Rifano, A. T. S. 2021. Justice and critical mineral development in Indonesia and across ASEAN. *The Extractives Industry and Society*, 8 (1), 355–362.

2. Heffron R. J. and McCauley, D. 2018. What is the 'Just Transition'? *Geoforum*, 88, 74–77.
3. Heffron, R. J. 2020. The Role of Justice in Developing Critical Minerals. *The Extractives Industry and Society*, 7.
4. Heffron, R. J. & McCauley, D. 2017. The concept of energy justice across the disciplines. *Energy Policy*, 105, 658–667.
5. The case study is redrafted from an article utilising the JUST Framework to explore critical mineral development, and for that the authors expresses thanks to the publishers and journal The Extractive Industry and Society—for more see: Heffron, R. J. 2020. The Role of Justice in Developing Critical Minerals. *The Extractives Industry and Society*, 7 (3), 855–863.
6. United Nations (UN). 2016. Mapping Mining to the SDGs: An Atlas. Available at: https://www.undp.org/content/undp/en/home/librarypage/poverty-reduction/mapping-mining-to-the-sdgs%2D%2Dan-atlas.html (Last accessed 1 September 2021).
7. See the following: (1) Akong, C. 2020. Reframing matter: Towards a material discursive framework for Africa's minerals. *The Extractive Industries and Society*, 7 (2) 461–469; (2) Andersson, P. 2020. Chinese assessments of "critical" and "strategic" raw materials: Concepts, categories, policies, and implications. *The Extractive Industries and Society*, 7 (1) 127–137; (3) Lee, J., Bazilian, M., Sovacool, B., Hund, K., Jowitt, S. M., Nguyen, T. P., Månberger, A., Kah, M., Greene, S., Galeazzi, C., Awuah-Offei, K., Moats, M., Tilton, J. and Kukoda, S. 2020. Reviewing the material and metal security of low-carbon energy transitions. *Renewable and Sustainable Energy Reviews*, 124, 109789; (4) Benjamin K. Sovacool, Saleem H. Ali, Morgan Bazilian, Ben Radley, Benoit Nemery, Julia Okatz, and Dustin Mulvaney. 2020. Sustainable minerals and metals for a low-carbon future. *Science*, 367 (6473), 30–33; (5) Bazilian, M. D. 2018. The mineral foundation of the energy transition. *The Extractive Industries and Society*, 5 (1) 93–97; (6) He, Y. 2018. The trade-security nexus and U.S. policy making in critical minerals. *Resources Policy*, 59, 238–249; (7) Coulomb, R., Dietz, S., Godunova, M. and Bligaard Nielsen, T. 2015. Critical minerals today and in 2030: an analysis of OECD countries. Prepared for the OECD by (Grantham Research Institute on Climate Change and the Environment at the London School of Economics and Political Science (LSE), London, U.K.). Available at: http://www.lse.ac.uk/GranthamInstitute/wp-content/uploads/2015/11/OECD_minerals_paper_Coulomb_et_al.pdf (Last accessed 1 September 2021); and (8) Viebahn, P., Soukup, O., Samadi, S., Teubler, J., Wiesen, K., Ritthoff, M. 2015. Assessing the need for critical minerals to shift the German energy system towards a high proportion of renewables. *Renewable and Sustainable Energy Reviews*, 49, 655–671.

8. European Commission. 2008. The raw materials initiative—meeting our critical needs for growth and jobs in Europe. COM (2008) 699. Available at: https://eur-lex.europa.eu/legal-content/EN/TXT/PDF/?uri=CELEX:52008DC0699&from=EN (Last accessed 1 September 2021).
9. McCellan, B.C., Corder, G. D., and Ali, S. H. 2013. Sustainability of Rare Earths—An Overview of the State of Knowledge Minerals 2013, 3, 304–317.
10. Heffron, R. J. 2018a. The application of distributive justice to energy taxation utilising sovereign wealth funds. *Energy Policy*, 122, 649–654.
11. United Nations (UN). 2019. Assessing the Rule of Law in Public Administration: The Mining Sector. Available at: https://www.undp.org/content/undp/en/home/librarypage/poverty-reduction/environmental-governance/ROL-MiningSector.html (Last accessed 1 September 2021).
12. United Nations (UN). 2018. A Guide for Governments and Partners to Integrate Environment and Human Rights into the Governance of the Mining Sector. Available at: https://www.undp.org/content/undp/en/home/librarypage/poverty-reduction/environmental-governance/extracting-good-practices%2D%2Da-guide-for-governments-and-partners-.html (Last accessed 1 September 2021).
13. Ali Saleem H., Damien Giurco, Nicholas Arndt, Edmund Nickless, Graham Brown, Alecos Demetriades, Ray Durrheim, and others. 2017. Mineral Supply for Sustainable Development Requires Resource Governance. *Nature* 543 (March 16), 367–372.
14. Vidal, Olivier, Bruno Goffé, and Nicholas Arndt. 2013. Metals for a Low-Carbon Society. *Nature Geoscience* 6 (October): 894–896.
15. Benjamin K. Sovacool, Saleem H. Ali, Morgan Bazilian, Ben Radley, Benoit Nemery, Julia Okatz, and Dustin Mulvaney. 2020. Sustainable minerals and metals for a low-carbon future. *Science*, 367 (6473) 30–33.
16. Lee, J., Bazilian, M., Sovacool, B., Hund, K., Jowitt, S. M., Nguyen, T. P., Månberger, A., Kah, M., Greene, S., Galeazzi, C., Awuah-Offei, K., Moats, M., Tilton, J. and Kukoda, S. 2020. Reviewing the material and metal security of low-carbon energy transitions. *Renewable and Sustainable Energy Reviews*, 124, 109789.
17. Heffron, R. J. 2018a. The application of distributive justice to energy taxation utilising sovereign wealth funds. *Energy Policy*, 122, 649–654.
18. Heffron, R. J. and Talus. K. 2016. The Evolution of Energy Law and Energy Jurisprudence: Insights for Energy Analysts and Researchers. *Energy Research and Social Science*, 19, 1–10.
19. See: (1) Stiglitz, J. E. 2012. *The Price of Inequality*. Penguin Books: London, UK; (2) Atkinson, A. B. 2015. *Inequality: What can be done?* MA, US: Harvard University Press; (3) Piketty, T. 2015. *The Economics of Inequality* (translated by Goldhammer, A.). Belknap Press of Harvard

University Press: MA, US; and (4) Scheidel, W. 2017. *The Great Leveler*. Princeton University Press: NJ, US.
20. Piketty, T. 2020. Capital and Ideology (translated by Goldhammer, A.). Belknap Press of Harvard University Press: MA, US.
21. Piketty, T. 2020. Capital and Ideology (translated by Goldhammer, A.). Belknap Press of Harvard University Press: MA, US.
22. Duff, D., 2017. Tax Policy and the Virtuous Sovereign: Dworkinian Equality and Redistributive Taxation, (p.167–189). In: Bhandari, M. (Ed.), Philosophical Foundations of Tax Law 2017 Oxford University Press, Oxford, UK.
23. Gribnau, H., Vording, H., 2017. The birth of tax law as a legal discipline. In: Harris, P., de
24. Extractive Industries Transparency Initiative. 2020. Available at: https://eiti.org/ (Last accessed 1 September 2021).
25. (1) The International Institute for Sustainable Development (IISD) and the Organisation for Economic Co-operation and Development (OECD). 2018a. Limiting the Impact of Excessive Interest Deductions on Mining Revenue. OECD: Paris, France; (2) The International Institute for Sustainable Development (IISD) and the Organisation for Economic Co-operation and Development (OECD). 2018b. Tax Incentives in Mining: Minimising Risks to Revenue. OECD: Paris, France; and (3) The International Institute for Sustainable Development (IISD) and the Organisation for Economic Co-operation and Development (OECD). 2018c. Monitoring the Value of Mineral Exports: Policy Options for Governments. OECD: Paris, France.
26. (1) Nogrady, B. 2019 (11 February). Landmark Australian ruling rejects coal mine over global warming, *Nature*, Available at: https://www.nature.com/articles/d41586-019-00545-8 (Last accessed 1 September 2021); and (2) Herbling, D. 2019 (26 June). Kenya cancels environment license of $2 billion coal-power plant', *Bloomberg*, Available at: https://www.bloomberg.com/news/articles/2019-06-26/kenya-cancels-environment-license-of-2-billion-coal-power-plant (Last accessed 1 September 2021).
27. Equator Principles. 2020. The Equator Principles. Available at: https://equator-principles.com/about/ (Last accessed 1 September 2021).
28. Heffron, R. J. & McCauley, D. 2017. The concept of energy justice across the disciplines. *Energy Policy*, 105, 658–667.
29. See: (1) Gallucci, M. 2016 (14 January). When a Coal Company Goes Bankrupt, Who Is Left to Clean up the Mess?' *International Business Times*. Available at: http://www.ibtimes.com/when-coal-company-goes-bankrupt-who-left-clean-mess-2264097 (Last accessed 1 September 2021); (2) Miller, C. G. 2005. Financial Assurance for Mine Closure and

Reclamation. Available: https://www.icmm.com/document/282 (Last accessed 1 September 2021); and (3) Robertson, J. 2016 (28 January). Coal Giants Abandon Unprofitable Mines, Leaving Rehabilitation under Threat. *The Guardian*. Available at: https://www.theguardian.com/environment/2016/jan/29/coal-giants-abandon-unprofitable-mines-leaving-rehabilitation-under-threat (Last accessed 1 September 2021).

30. Heffron, R. J. et al. 2018a. The emergence of the 'social licence to operate' in the extractive industries? *Resources Policy, Available early-access*. https://doi.org/10.1016/j.resourpol.2018.09.012.
31. Heffron, R. J. 2018c. Energy law for decommissioning in the energy sector in the 21st century. *Journal of World Energy Law & Business*, 11 (3), 189–195.
32. See (1) ABC Australia (Sue Lannin). 2015 (18 December). China Economist Warns Major Miners May Collapse in 2016. Available at: http://www.abc.net.au/news/2015-12-17/china-economist-warns-that-iron-ore-miners-will-collapse/7037802 (Last accessed 1 September 2021); and (2) Bloomberg, 2016 (3 December—Thomas Biesheuvel, Jesse Riseborough & Agnieszka De Sousa). Why Bankruptcy Might Be the Mining Industry's Last Best Hope. Available at: http://www.bloomberg.com/news/articles/2015-12-03/why-bankruptcy-might-be-the-mining-industry-s-last-best-hope (Last accessed 1 September 2021).
33. McCauley, D., Heffron, R. J. Stephan, H. and Jenkins, K. 2013. Advancing Energy Justice: The triumvirate of tenets. *International Energy Law Review*, 32 (3), 107–110.
34. United Nations Declaration of the Rights of Indigenous Peoples (UNDRIP). 2007. Available at: https://www.un.org/development/desa/indigenouspeoples/declaration-on-the-rights-of-indigenous-peoples.html (Last accessed 1 September 2021).
35. Phillips, J. S. 2015. The rights of indigenous peoples under international law. *Global Bioethics*, 26 (2), 120–127.
36. UN Economic and Social Council, 2012. An Analysis on the Duty of the State to Protect Indigenous Peoples Affected by Transnational Corporations and Other Business Enterprises. E/C.19/2012/3. Available at: https://www.un.org/esa/socdev/unpfii/documents/2012/session-11-e-c19-2012-3.pdf (Last accessed 1 September 2021).
37. Heffron, R. J. et al. 2018a. The emergence of the 'social licence to operate' in the extractive industries? *Resources Policy, Available early-access*. https://doi.org/10.1016/j.resourpol.2018.09.012.
38. Heffron, R. J. et al. 2018a. The emergence of the 'social licence to operate' in the extractive industries? *Resources Policy, Available early-access*. https://doi.org/10.1016/j.resourpol.2018.09.012.

39. International Bar Association (IBA). 2016. *IBA Practical Guide on Business and Human Rights for Business Lawyers.* The International Bar Association: London, UK.
40. International Bar Association (Human Rights Institute). 2013. *Tax Abuses, Poverty and Human Rights.* IBA: London, UK. Available at: https://www.ibanet.org/Article/NewDetail.aspx?ArticleUid=4A0CF930-A0D1-4784-8D09-F588DCDDFEA4 (Last accessed 1 September 2021). It should be noted also that this book is available to download freely.
41. United Nations Economic and Social Council (UNESC). 2017 (11 August). *Committee on Economic and Social and Cultural Rights—General comment No. 24 (2017) on State obligations under the International Covenant on Economic, Social and Cultural Rights in the context of business activities.* Available at: http://docstore.ohchr.org/SelfServices/FilesHandler.ashx?enc=4slQ6QSmlBEDzFEovLCuW1a0Szab0oXTdImnsJZZVQcIMOuuG4TpS9jwIhCJcXiuZ1yrkMD%2FSj8YF%2BSXo4mYx7Y%2F3L3zvM2zSUbw6ujlnCawQrJx3hlK8Odka6DUwG3Y (Last accessed 1 September 2021).
42. Heffron, R. J. 2018b. Mining Disputes. In Scherer, M. 2018. (Editor). International Arbitration in the Energy Sector. Oxford University Press: Oxford, UK.
43. Canadian Ombudsperson for Responsible Enterprise (CORE). 2020. Office of the Canadian Ombudsperson for Responsible Enterprise. Available at: https://core-ombuds.canada.ca/core_ombuds-cre_ombuds/index.aspx?lang=eng (Last accessed 1 September 2021).
44. Volterra Fietta. 2018. The UK and Canada are taking steps to regulate the area of business and Human Rights ("BHR"). Available at: https://www.volterrafietta.com/the-uk-and-canada-are-taking-steps-to-regulate-the-area-of-business-and-human-rights-bhr/ (Last accessed 1 September 2021).
45. UNFCC. 2020. Paris Agreement—Status of Ratification. Available at: https://unfccc.int/process/the-paris-agreement/status-of-ratification (Last accessed 1 September 2021).
46. UN. 2020. Convention on Environmental Impact Assessment in a Transboundary Context Available at: https://treaties.un.org/Pages/ViewDetails.aspx?src=TREATY&mtdsg_no=XXVII-4&chapter=27&lang=en#1 (Last accessed 1 September 2021).
47. Van Bockstael, S. 2018. The emergence of conflict-free, ethical, and Fair Trade mineral supply chain certification systems: A brief introduction. *The Extractive Industries and Society*, 5 (1) 52–55.
48. See: (1) Agnew, J. 2011. Space and Place. In J. Agnew and D. Livingstone (eds.) *Handbook of Geographical Knowledge.* London: Sage; and (2) Gorter, C. and Nijkamp, P. 2001. Location Theory. International Encyclopedia of the Social & Behavioral Sciences 2001, Pages 9013–9019, Elsevier.

49. Amnesty International, 2017. Time to Recharge: Corporate Action and Inaction to Tackle Abuses in the Cobalt Supply Chain. Available at: https://www.amnesty.org/download/Documents/AFR6273952017ENGLISH.PDF (Last accessed 1 September 2021).
50. Amnesty International, 2017. Time to Recharge: Corporate Action and Inaction to Tackle Abuses in the Cobalt Supply Chain. Available at: https://www.amnesty.org/download/Documents/AFR6273952017ENGLISH.PDF (Last accessed 1 September 2021).
51. Heffron, R. J. et al. 2018a. The emergence of the 'social licence to operate' in the extractive industries? Resources Policy, Available early-access. https://doi.org/10.1016/j.resourpol.2018.09.012.
52. (1) The International Institute for Sustainable Development (IISD) and the Organisation for Economic Co-operation and Development (OECD). 2018a. Limiting the Impact of Excessive Interest Deductions on Mining Revenue. OECD: Paris, France; (2) The International Institute for Sustainable Development (IISD) and the Organisation for Economic Co-operation and Development (OECD). 2018b. Tax Incentives in Mining: Minimising Risks to Revenue. OECD: Paris, France; and (3) The International Institute for Sustainable Development (IISD) and the Organisation for Economic Co-operation and Development (OECD). 2018c. Monitoring the Value of Mineral Exports: Policy Options for Governments. OECD: Paris, France.
53. Figueres, C. et al. 2017. Three Years to Safeguard our Climate. *Nature*. 546 (7660), 593–595. (29 June 2017).
54. Ali Saleem H., Damien Giurco, Nicholas Arndt, Edmund Nickless, Graham Brown, Alecos Demetriades, Ray Durrheim, and others. 2017. "Mineral Supply for Sustainable Development Requires Resource Governance." *Nature* 543 (March 16), 367–372.
55. UK Government. 2016. *Coal generation in Great Britain: The pathway to a low-carbon future*. Available at: https://www.gov.uk/government/consultations/coal-generation-in-great-britain-the-pathway-to-a-low-carbon-future (Last accessed 1 September 2021).
56. International Institute for Sustainable Development. (IISD) (Kosolapova, E.). 2019 (24 September). 77 Countries, 100+ Cities Commit to Net Zero Carbon Emissions by 2050 at Climate Summit. Available at: http://sdg.iisd.org/news/77-countries-100-cities-commit-to-net-zero-carbon-emissions-by-2050-at-climate-summit/ (Last accessed 1 September 2021).
57. Heffron, R. J., Ronne, A., Bradbrook, A., Tomain, J. P. and Talus, K. 2018b. A Treatise for Energy Law. *Journal of World Energy Law & Business*, 11 (1), 34–48.

CHAPTER 4

The Advance of Just Transition Commissions

Abstract Some countries have already begun to realise the enormity of this challenge of a *just transition* to a *low-carbon economy*, and they are putting in place the first legislative steps to achieve it. This is through the creation of a *just transition commission*, and there are currently ten forms of such a commission in the following jurisdictions: Canada, Germany, Scotland, Australia, Ireland, New Zealand, the USA (Appalachia), South Africa, the EU and the UN. Such a commission will provide expert advice on the ways to achieve a just transition and also will monitor the effects of existing laws and policies to ensure they contribute to the delivery of a just transition. This chapter will examine these early legal steps in that selection of countries.

Keywords Just transition • Just transition commissions • Just Framework • Climate timelines

---

It is with express thanks to my Research Assistant John O'Boyle who worked on this chapter in 2021 with me and responded to all direction in assisting with the different drafts of the chapter.

R. J. Heffron, *Achieving a Just Transition to a Low-Carbon Economy*, https://doi.org/10.1007/978-3-030-89460-3_4

4.1 Introduction: Background to the Rise of Just Transition Commissions

There is a range of perspectives of what the just transition to a low-carbon economy is. For different communities of people, it means different things. There is a research, a practitioner (private sector), government and policy-making perspective and justice, technological and the general public perspective. These different dimensions of the just transition to a low-carbon economy all need to align in order for it to be realised. The very essence of this paper is about how this happens in the form of a new development, the Just Transition Commission (hereafter referred to as the "JTC").

Some countries have already begun to realise the enormity of this challenge of a *just transition* to a *low-carbon economy*, and they are putting in place the first legislative steps to achieve it. This is through the creation of a *Just Transition Commission*, and there are currently ten forms of such a commission in the following jurisdictions: Canada, Germany, Scotland, Australia, Ireland, New Zealand, the US (Appalachia), South Africa, the EU and one international institution. Such a commission will provide expert advice on the ways to achieve a just transition and also will monitor the effects of existing laws and policies to ensure they contribute to the delivery of a just transition.

Such a step as establishing legal entities such as a commission to oversee the just transition is a significant step forward. In essence, it demonstrates the way forward that governments see of how to deliver the just transition which is to see it as the new social contract. To deliver such a social contract needs collaboration from all, not just labour unions but also the entire communities of researchers and practitioners from across the areas of energy, environment, climate change and sustainability. As much as there needs to be a societal shift in the way economies operate to move towards a low-carbon economy, so too there needs to be a shift in thinking concerning how the just transition is researched and advanced.

As stated previously the major legal tool to ensure there is a just transition is to create a Just Transition Commission (JTC). Ten countries and entities are examined, here where a form of a JTC is in the process of or has already been introduced, and these are Canada, the United Nations, Germany, Scotland, Australia, Ireland, New Zealand, the European Union, the US and South Africa.

Key questions concern the formulation of these JTCs. In legal terms these JTCs would be a form of a new public administrative unit (i.e.

attached to a ministerial department such as energy, labour, etc. or spanning a few departments). There is a distinct area of legal scholarship known as public administrative law, and in this context from literature in this area, there are a number of key questions to ask about a JTC, and these would be:

1. What are the powers of the JTC?
2. What persons and/or who is represented on the JTC?
3. What is the lifespan and reporting periods for the JTC?
4. What knowledge is utilised in decision-making for the JTC?

These questions are important as they will directly impact on how effective a JTC can be in a particular country. Early research into JTCs indicates that countries are going for a number of options, but there are two emerging solutions which can be identified as an external JTC and an internal JTC. An external JTC is now in evidence in Germany (2018) and in Scotland (2019), and this is where the majority of the membership of the JTC consists of external appointments—that is, they represent different stakeholders in society in terms of the development of the just transition to a low-carbon economy. The internal JTC (New Zealand, 2019) in contrast consists of civil servants who then engage with stakeholders.

The issue of the powers of a JTC remains undefined. And there is research needed into exploring which type of JTC, the internally or externally oriented one, will have more power. As of yet, the powers of the JTC are slowly beginning to emerge, and this is in essence a "live" research issue. Indications are that this is an area that governments are approaching with caution; it is politically sensitive and also requires cooperation from multiple government ministries. However, if there is success, a JTC could have a transformative effect in the development of low-carbon economies worldwide. Alongside other scholars worldwide, we hope to answer the questions highlighted above as these issues continue to develop.

4.2 Brief Examination of the Just Transition Commissions

4.2.1 Australia

Following the closure of the Hazelwood coal power station in 2017[1] and the government intervention at regional[2] and federal levels[3] to avert the crisis faced by hundreds of workers the station employed, conversation within Australian union circles focussed on how a national strategy could take shape in order to prevent workers like those employed by Hazelwood from finding themselves in a similar situation as a result of the global transition in the energy markets away from fossil fuels and towards renewable energy.[4]

In 2016 the Australian Senate referred the subject of retiring coal-fired power stations to the Environment and Communications References Committee for a final report by 2017. Upon the publication of the final report in 2017, the ACTU "recommended that the Australian Government establish a national independent statutory authority, named **Energy Transition Australia (ETA)**, within the environment and energy portfolio, 'to navigate the transition to a clean energy economy'".[5]

The ACTU went on to explain that the role of the ETA would be to

> oversee a planned and orderly closure of Australia's coal fired power stations…manage an industry-wide multi-employer pooling and redeployment scheme, where existing workers would have an opportunity to be redeployed to remaining power stations or low-emissions generators…develop a labour adjustment package to support workers obtain new decent and secure jobs, including by providing funding for workers to access job assistance support, retraining, early retirement and travel and relocation assistance.[6]

At the time of writing in 2021, the ETA as envisioned by the ACTU in the 2017 report has not been established.

4.2.2 *Canada*

As part of their climate action plan, the Canadian government in 2018 committed to phasing out coal-fired electricity by 2030.[7] Given the implications of this commitment for workers and communities supported by the coal industry, the Canadian minister for environment and climate change established the **Task Force on the Just Transition for Canadian Coal Power Workers and Communities**.[8] At the end of 2018 the Task Force produced its final report for the minister.

In its report, the Task Force provided ten recommendations:[9]

- Develop, communicate, implement, monitor, evaluate and publicly report on a just transition plan for the coal phase-out, championed by a lead minister to oversee and report on progress.
- Include provisions for just transition in federal environmental and labour legislation and regulations, as well as relevant intergovernmental agreements.

- Establish a targeted, long-term research fund for studying the impact of the coal phase-out and the transition to a low-carbon economy.
- Fund the establishment and operation of locally driven transition centres in affected communities.
- Create a pension bridging programme for workers who will retire earlier than planned due to the coal phase-out.
- Create a detailed and publicly available inventory with labour market information pertaining to coal workers, such as skills profiles, demographics, locations and current and potential employers.
- Create a comprehensive funding programme for workers staying in the labour market to address their needs across the stages of securing a new job, including income support, education and skills building, re-employment and mobility.
- Identify, prioritise and fund local infrastructure projects in affected communities.
- Establish a dedicated, comprehensive, inclusive and flexible just transition funding programme for affected communities.
- Meet directly with affected communities to learn about their local priorities and to connect them with federal programmes that could support their goals.

4.2.3 *The European Union*

In 2017 at an event entitled No Region Left Behind, the EU launched the **Platform For Coal Regions in Transition**, the purpose of which was to "facilitate the development of projects and long-term strategies in coal regions, with the aim of kick-starting the transition process and responding to environmental and social challenges".[10] As well as that, there it was also intended "to boost the clean energy transition by bringing more focus to social fairness, structural transformation, new skills and financing for the real economy".[11]

The way that the platform is intended to operate is through the holding of regular meetings, the establishment of pilot projects in a number of EU regions and through the delivering of reports and toolkits.[12] In the years succeeding the launch of the platform, the European Commission has launched a broader just transition strategy known as the **Just Transition Platform**, which appears to be an attempt to streamline the range of transition-related initiatives that the EU has launched, with the commission website remarking that the "…Just Transition Platform aims to assist

EU countries and regions to unlock the support available through the Just Transition Mechanism. This platform will provide a single access point for support and knowledge related to the just transition".[13]

The commission website goes on to explain that "The Just Transition Platform will build on and expand the work of the existing **Initiative for Coal Regions in Transition**, which already supports fossil fuel producing regions across the EU in achieving a just transition through tailored, needs-oriented assistance and capacity-building".[14]

4.2.4 *Germany*

In November 2016 the German government adopted its climate action plan 2050 in which the country aims to reduce its greenhouse gas emissions and specify its climate targets with the aim of becoming greenhouse gas emission neutral by 2050.[15] The plans to achieve this long-term target as well as the short-term target of reducing emissions by 55% by 2030 compared to 1990 levels[16] involved restructuring the energy sector, a transition which would have a significant impact on some German regions.

In order to ensure the communities where this restructuring would take place did not suffer in terms of structural development and employment as a result of the restructuring, the German Federal cabinet launched the **Commission on Growth, Structural Change and Employment** in 2018.[17] The commission was tasked with submitting initial social and economic policy recommendations for coal regions including concrete measures for new, future-proof jobs in lignite regions affected by the structural changes and for financial security in October 2018, followed by recommendations for closing the gap to Germany's 2020 climate target "as much as possible".[18] The commission was then due to publish its final report for the government in 2018 which included a recommended end date for coal in Germany.

The Commission published the report, citing 2038 as its recommended year for the cessation of coal in Germany.[19]

4.2.5 *International: The United Nations*

The Paris Climate Agreement of 2015 committed 193 countries from around the world to a legally binding agreement to phase out fossil fuels and limit global warming to 1.5 degrees Celsius while at the same time explicitly mentioning the need for a just transition.[20] While mentioning the need for a just transition, the Paris Agreement did not provide guidance as to how this was to be achieved, though this guidance gap was subsequently filled by the International Labour Organisation, who adopted **guidelines** for a just transition for environmentally sustainable economies and societies for all.[21] The guidelines provide a policy framework and a practical tool to ensure that national and international attempts to tackle climate change advance employment creation targets and social justice and encourage just transitions for workforces, companies and communities.

The guidelines did not explicitly call for the forming of a just transition commission, acknowledging that each country had unique economic, social and environmental circumstances making a broad one-size-fits-all commission difficult.[22] Its guidelines do however suggest that the ILO believes the concept of a just transition should be woven throughout all levels of government and all government departments as a defining feature of policy formation.[23]

As of the time of writing, there still is no international just transition commission of the scale, level of authority and involvement from countries all over the world that would be needed to make the just transition; however COP 26 is due to take place this year so it shall be important to check the outcome of those talks for any pertinent developments in regard to the international just transition.[24]

4.2.6 Ireland

The Irish government tried a number of different initiatives to achieve its EU and international obligations to reduce emissions. Following years of different climate initiatives, in 2019 the Government of Ireland published its Climate Action Plan which called for the establishment of a **Just Transition Review Group** within the National Economic and Social Council.[25] Through this group the NESC will "review the ongoing transition and identify specific transition needs among cohorts of workers, enterprises, communities and specific groups of people. It will collaborate and engage with a wide range of stakeholders and will work closely with the NDCA and SDGs Stakeholder Forum."[26]

Following the establishment of the review group which advised the NESC on aspects of the just transition, areas of concern and negative impacts and how to mitigate them, the NESC completed its research and published its advice to the government in the form of a document entitled

"Addressing Employment Vulnerability as Part of a Just Transition in Ireland".[27] The paper states, "The transitions to a low-carbon, more technological Ireland are underway. The Government has correctly sought to play its part in mitigating any negative impacts these changes may have on employment. The Council has identified 12 key recommendations to be taken forward by government",[28] before going on to state their recommendations under four broad headings of "Continuous, pre-emptive workforce development…Building resilient enterprises…Delivering high-impact targeted funding to support transition…Making it happen".[29]

4.2.7 New Zealand

New Zealand has used a number of different initiatives in its aim to reach net zero by 2050, with its establishment initially of an interim climate change committee being the closest thing that the government had to a just transition commission.[30] This interim commission's purpose was to develop evidence and analysis on key issues for climate change policy in New Zealand, in particular agriculture and renewable electricity, whilst also taking into account the importance of just transition, in particular the likely impact of changes in policy on people, communities, Maori (indigenous Polynesian people of New Zealand) and other groups.[31]

The ICCC handed in its final report in November 2019 at which point it was replaced with the **Climate Change Commission**.[32] The Climate Change Commission's website claims the crown entity's purpose was "to provide independent, evidence-based advice to Government to help Aotearoa transition to a climate-resilient and low emissions future".[33] The Commission is also "tasked with monitoring and reporting on the government's progress towards meeting the 2050 target, and on progress against emissions budgets and reduction plans. We also monitor how national adaptation plans are implemented".[34]

In January 2021 the commission released its first report which was critical of the approach taken by the New Zealand government and encouraged more rapid action to meet its emission targets.[35] However, it also emphasised the need for a more accelerated transition to be a just one: "Aotearoa must have an equitable and fair transition to a low emissions economy and society with benefits widely shared".[36]

4.2.8 *Scotland*

In September 2017, the Scottish Government committed to establish the **Just Transition Commission** which aims to eliminate greenhouse gas emissions while promoting greater social justice and environmental sustainability through advising ministers.[37] This commission's purpose was "to advise Scottish Ministers on how to apply Just Transition principles to Scotland".[38]

Within two years of its inaugural meeting in January 2019, "the Commission will provide a written report to Scottish Ministers that provides practical, realistic, affordable recommendations for action".[39] The advice would seek to give ministers the necessary knowledge to maximise the social and economic opportunities afforded by a net zero by 2045 economy whilst also ensuring ministers "understand and mitigate risks that could arise in relation to regional cohesion, equalities, poverty (including fuel poverty), and a sustainable and inclusive labour market".[40] On February 27, 2020, the commission published its interim report, followed by the final report on March 23, 2021, in a report titled "A National Mission for a fairer, greener Scotland".[41]

In its final report, the commission summarised its call to action for the incoming New Scottish Government with four messages: "Pursue an orderly, managed transition to netzero that creates benefits and opportunities for people across Scotland. Delivery of this must be a national mission…Equip people with the skills and education they need to benefit from the transition…Empower and invigorate our communities and strengthen local economies…Share benefits widely and ensure burdens are distributed on the basis of ability to pay".[42]

4.2.9 *South Africa*

In 2010 the National Planning Commission was established to develop a long-term vision and strategic plan for South Africa. The main objective of the commission is to rally the nation around a common set of objectives and priorities to drive development over the longer term. The commission advises government on cross-cutting issues that influence the long-term development of South Africa.[43]

Up until late 2020, the NPC dealt with significant amounts of South Africa's energy transition approach and strategy. However following a Presidential Jobs Summit held in late 2018, social partners agreed that a statutory body needed to be formed "to coordinate and oversee the just transition towards a low-carbon, inclusive, climate change resilient economy and society".[44] This led to the establishment of the **Presidential Climate Change Coordinating Commission**.

The commission is tasked "with advising on South Africa's climate change response. This includes mitigation and adaptation to climate change and its associated impacts. It will furthermore provide independent monitoring and review of South Africa's progress in meeting its emissions reduction and adaptation goals".[45]

Furthermore "Under the Commission's Terms of Reference it will advise on and facilitate a common understanding of a just transition, cognisant of the socio-economic, environmental and technological implications of climate change. This covers adaptation, mitigation as well as means of implementation. It will also provide a platform for the engagement of key stakeholders on the National Employment Vulnerability Assessment (NEVA) and Sector Job Resilient Plans (SJRPs) and ensure reporting of progress towards the implementation of the Sector Job Resilient Plans".[46]

4.2.10 The USA

The **Appalachian Regional Commission** (ARC) is a regional economic development agency that represents a partnership of federal, state and local government. It was established by an act of Congress in 1965;[47] ARC is composed of the governors of the 13 Appalachian states and a federal co-chair. The Appalachian Region includes all of West Virginia and parts of 12 other states: Alabama, Georgia, Kentucky, Maryland, Mississippi, New York, North Carolina, Ohio, Pennsylvania, South Carolina, Tennessee and Virginia.[48]

The Appalachian Region's economy, once highly dependent on mining, heavy industry and forestry, has recently become more diversified. The region includes 420 counties in 13 states. ARC receives the federal funds to help the region to transition from a coal-dominated mono-economy to a more robust economic environment.[49] Since March 2016 when the ARC first announced funding availability through the power initiative, it has received more than $280 million in funding request to revitalise the region's economy.[50] Its work continues at the time of writing despite the previous President Donald Trump's effort to eliminate ARCs funding in 2018.[51]

4.3 Just Transition Commissions Analysed with the JUST Framework

In this section, Just Transition Commissions are analysed with the JUST Framework. This is what follows for the remainder of this section.

4.3.1 *Justice and Its Elements: Distributive, Procedural and Restorative Justice*

4.3.1.1 Procedural Justice

From a procedural perspective, the purpose of all of the various commissions discussed above is to provide aid in facilitating a transition from the carbon-dependent societies which are common across the world today to lower-carbon or net zero societies, usually by 2050 at the latest. Whilst facilitation of these general goals is common across almost all of the commissions discussed previously, there can be significant divergence across the different commissions when it comes to the correct procedural approach for facilitating these goals.

Many of the commissions fulfil a purpose of advising their government on the best approach to achieving their climate change and emission reduction goals through engaging in a period of consultation following their establishment and working towards deadlines by which they usually have to have produced first an interim report, as can be seen in the cases of Scotland,[52] New Zealand[53] and Germany[54] before they submit a final report to the government on how best to facilitate a just transition to a low-carbon or carbon-neutral economy.

This procedural pattern of establishment, consultation, interim report(s) and final report can best be observed in the procedural approach taken by the governments of New Zealand, Germany and Scotland who at the time of writing have all had a final report from their respective commissions submitted, containing recommendations for transition to low-carbon economies within.[55,56,57] Whilst the three aforementioned commissions lead the rest in terms of procedural coherency (establishment, consultation, interim report(s), final report), other countries are also succeeding at adopting procedurally coherent approaches to helping facilitate just transitions, with some minor variations.

For example, Canada followed an approach which was much the same as the previously mentioned countries though they opted to skip the interim reports in favour of simply producing a final report.[58] Ireland on the other hand created their Just Transition Review Group as a subsidiary body of the National Environmental and Social Council which in essence assumed the just transition remit whilst establishing their own in-house review group which produced a report on the NESC's behalf, which the NESC then presented to the Irish government.[59]

Ireland additionally appointed Kieran Mulvey as a just transition commissioner in late 2019 in order to allay disquiet in the Irish midlands over peat-fired power station closures.[60] Mulvey produced his first report on the matter in mid-2020.[61] Whilst Ireland is different to Scotland in that their JTC was established a subsidiary body within the NESC rather than as its own separate body, as seen with Scotland, the final report of the Scottish JTC recommended for the appointment of a role very similar to Ireland's just transition commissioner. They recommended the incoming Scottish government "should make the Deputy First Minister, or at a minimum, a Cabinet Secretary responsible for a just transition to net-zero".[62]

Here then it is clear that there are two slight variations on the procedural patterns used by New Zealand, Germany and Scotland. While the ultimate outcomes are the same in that governments receive a report compiled by a JTC, the process vary slightly from the first three countries through omission of one step (Canada) and the JTC being established as a smaller body within an existing body, rather than being a separate entity (Ireland).

Elsewhere the procedural differences are more noticeable due to a variety of criteria. In some instances, such as Australia, the procedural coherency in facilitating a just transition is not there simply because in Australia's case a JTC has yet to be formed, and therefore there is not an existing procedure by which a JTC could go about fulfilling an objective of transitioning Australia towards being a low-carbon or carbon-neutral economy, though there has been an idea suggested for such a commission.[63] In South Africa's case, the procedural coherency of its approach to a just transition cannot be evaluated yet as the Presidential Climate Change Coordinating Commission was only formed late last year,[64] and so it is a few years behind the first five aforementioned countries.

The USA's Appalachian Regional Commission, whilst having some just transition-related aims and functions, is a body whose purposes are much broader than a just a just transition (unlike many of the previously mentioned bodies). This combined with its non-national character (ARC is a regional development agency whose remit covers 13 states[65]) makes it difficult to make a procedural comparison between ARC and the other

JTCs mentioned. Elsewhere the EU's Just Transition Platform, rather than primarily being a body which consults and provides reports to a national government (though it also serves those functions), acts as a funding and capacity building body to an international 27 country political and economic union, which concentrates its functions towards specific areas of specific countries in that union.[66]

Finally when it comes to international bodies and the ILO, the ILO has simply provided guidelines which provide advice to all countries around the world for how to go about taking procedural steps to making a just transition whilst being careful to make clear that an international JTC would be hard to make work due to the differences across a range of criteria that exist in each country.[67] For this very reason, attempting to facilitate a just transition through the establishment of an international JTC would be very challenging from a procedural perspective, due to the economic and social development differences which exist around the world.

In conclusion, a procedural analysis of the different JTCs seems to suggest a pattern exists between procedural coherency in achieving a just transition and the scale of the transition in question. For the most part, national governments of individual countries adopt similar procedural approaches with some minor procedural differences. International or large-scale JTCs such as we see in the USA, the EU and with international bodies such as the ILO deviate from the national approaches of the other countries, possibly on account of the scale of the transitions involved which makes a straightforward procedurally coherent transition difficult. Australia is the only outlier in not having an existing body which serves the functions of a JTC.

4.3.1.2 Distributive Justice

From a distributive perspective, a just transition to low-carbon/carbon-neutral societies is as much about tangible changes that people can recognise in their everyday lives as it is about changing energy sources and keeping the level of global heating low. A distributive analysis of the JTCs forces one to assess the extent to which the JTC in question shall generate new sustainable jobs to replace those which are lost, seek to compensate those who cannot have their job replaced, aim to provide retraining so that a labour force affected by the transition can retrain to re-enter the labour market and invest money in regions which have suffered from deprivation and any other tangible, quantifiable benefits. If we look firstly at the JTCs which have completed final reports providing transition recommendations to their respective governments, there are several distributive benefits which can be noted.

In Canada, the JTCs final report came with ten recommendations, of which approximately eight could be regarded as being distributive benefits: "establish a targeted, long-term research fund... the establishment and operation of locally-driven transition centres in affected communities... a pension bridging program for workers who will retire earlier than planned... a detailed and publicly available inventory with labour market information pertaining to coal workers... a comprehensive funding program for workers staying in the labour market to address their needs... identify, prioritize, and fund local infrastructure projects.... meet directly with affected communities to learn about their local priorities and to connect them with federal programs".[68]

Aside from two of the ten recommendations which focus more upon establishing government-level just transition strategies, the recommendations which emerge from Canadas JTC lean heavily into making the transition a distributive one for Canadians where affected communities and workers shall have access to a diverse range of programmes, funding and access to relevant government figures to ensure that Canadians affected by the transition have the necessary resources to mitigate possible negative impacts of the transition. The Canadian reports recommendations have less of an emphasis on the creation of good sustainable jobs but the focus on mitigation is comprehensive and impressive.

The German JTCs final report features an entire section called "Prospects for existing, new and future proof-jobs".[69] In this section the commission adopts a region-by-region approach to the transition and makes region-specific recommendations as to how adaption to the transition can have region-specific plans to mitigate the negatives and focus on each region's potential: "the commission gained a grass roots impression of the challenges which the regions are already facing and the effects which the premature termination of coal-fired power generation can have in the regions. In addition, the commission gathered information on the existing potential in the regions to face these challenges successfully and to benefit from the opportunities which arise from the structural transition".[70]

Hence, with a focus upon infrastructure development, funding and monitoring bodies, the German JTC report provides an impressive region-specific approach to transitioning that gives government ministers specific guidance for fostering tangible distributive benefits in all affected regions

> Identifiable and sufficient establishment of new jobs and new value creation in the regions to achieve equivalent replacement for the decrease in jobs…An immediate action structural policy programme and an immediate action programme for entrepreneurial investments…Creation of customised programmes for the mining areas to promote joint research and development by science and business, and initial location projects for research institutes (in 2023); … creation of participative bodies to ensure that employers, employee representatives, business and civil society stakeholders in the locality are institutionally involved in approvals for subsidy projects and the distribution of funds (in 2023).[71]

Elsewhere, the Irish report submitted by NESC hits many of the same notes as the German JTC report, though it could be argued that there is a slight air of vagueness and lack of ambition regarding their recommendations: "Increase Ireland's ambition for workforce development with higher targets for employee training, appropriate resourcing of key bodies, and incentives for employers to help workers attend programmes… Increase the momentum behind efforts to increase participation in lifelong learning, especially by those who are under-represented in this activity… Government…should take forward the recommendations of this report to provide more proactive supports for workers, improve enterprise resilience and help target finance to support transition".[72] The NESC's recommendations make no mention of creating good new sustainable jobs or establishing many new transition-focussed bodies aside from a Just Transition Review Group to start a societal dialogue on a just transition.[73]

Their focus is more so on insulating the labour force which already exists and adjusting those members who may be vulnerable to the realities a transition entails, rather than significantly expanding the labour force with good sustainable jobs or increasing the avenues through which a just transition can be a vehicle for positive social and economic transition through the establishment of bodies such as those proposed in Canada. On this basis, from a distributive point of analysis, the NESC offers one of the more conservative just transitions out of the different countries studied.

These three countries whose reports were completed and submitted simultaneously display the similarities and differences which can be revealed through a distributive analysis of different JTCs and what each of them offer (or refrain from offering) when it comes to tangible, quantifiable benefits for the people of their respective countries.

ARC in the USA, whilst a different body in many ways to the first three countries, does share a similarity with them in that it has a strong distributive focus, with the commission seeking to mitigate economic stagnation in the Appalachian Region for which it is responsible by undertaking the development of large infrastructure projects, such as the Appalachian Development Highway System and maintaining that whilst currently investing in the development of additional infrastructure and workforce development.[74] South Africa is light on details about what its JTC intends to offer from a distributive perspective so far, which is understandable to an extent given how recently it was formed, but it does refer to the JTC providing "a platform for the engagement of key stakeholders on the National Employment Vulnerability Assessment (NEVA) and Sector Job Resilient Plans (SJRPs) and ensure reporting of progress towards the implementation of the Sector Job Resilient Plans",[75] which vaguely hints at the JTC having a distributive agenda.

It could reasonably be predicted that an analysis of the remaining countries and international institutions JTC reports and goals would also reveal the same patterns of similarity and differences which can be seen here. Almost all will allude to jobs and transition-related bodies as well as funding, and there may be a region-specific focus, but analysis would also reveal differences of emphasis on different benefits and varying levels of specificity. This could be for a range of reasons, but would seem to corroborate the ILO's assessment that when it comes to just transitions, there is no one-size-fits-all approach.[76]

4.3.1.3 *Restorative Justice*

A restorative analysis of the different JTCs can examine the restorative credentials of each commission principally from two perspectives: to what extent is there a focus upon restoring land and some of the natural environment which has suffered damage during the preceding decades and to what extent are jobs and employment opportunities that have been lost or will be lost as a result of a just transition be restored via a just transition, though it is also important to note that region-specific restoration can also be detected in certain JTCs (this will be touched upon in more depth in the space and time section), whereas other JTCs focus upon conservation and mitigation over restoration.

The guidelines of the ILO possess both restorative elements of a just transition. When it comes to restoring the natural environment, the ILO explains that its guidelines aim to "promote cooperation...through initiatives...particularly with regard to low-carbon climate-resilient agriculture, renewable energy, restoration of natural resources and reforestation activities".[77]

As well as having environmentally restorative aspects, the ILO's guidelines also make reference to a just transition being a vehicle to restore job opportunities which have already been or shall be lost, through the just transition: while on the one hand the ILO acknowledges the challenges of a just transition from the perspective of job losses, explaining that "In the transition to environmentally sustainable economies and societies, the world of work... faces some major challenges, for example...economic restructuring, resulting in the displacement of workers and possible job losses";[78] it goes on to assert that "the world of work can benefit from some major opportunities, for example...net gains in total employment from realizing the potential to create significant numbers of additional decent jobs through investments into environmentally sustainable production and consumption and management of natural resources".[79] When it comes to the restorative benefits of a just transition then, the ILO is eager to emphasise the environmental and job-related benefits of the undertaking.

Elsewhere, the NESC's report for the Irish government could arguably be one of the stronger reports when subject to a restorative analysis. The NESC's report explains that "The Joint Committee viewed 'just transition' as exploring opportunities...creating new jobs in areas such as...peatland restoration...peatland restoration of degraded lands will require skilled ecologists",[80] linking environmental restoration with employment opportunities brought about by a just transition. The NESC's report, when quoting the climate change advisory council, even makes a specific reference to the idea of "restorative justice" in reference to a just transition, saying that "responding to climate change is an opportunity to engage in restorative justice and requires actively engaging vulnerable and underrepresented groups in terms of gender, ethnicity and socio-economic status while developing responses to climate change".[81] These references within the NESC's report demonstrate a recognition of the environmental and jobs related restorative benefits of a just transition and to some extent offsets the report's lacklustre distributive offerings.

The ACTU in Australia, whilst putting forward their argument for the establishment of an ETA/JTC, also touches upon the importance of the restorative potential of a just transition, arguing that, from a jobs perspective, "Taking action on climate change will create many new jobs.... Germany has managed to phase out its hard coal mines without a single forced redundancy as a result of significant government planning, investment and institutional support over a period of 2 decades"[82] before going on to explain that environmental restoration can also provide employment opportunities: "Comprehensive and funded mine and power station site rehabilitation plans which can provide a significant source of employment. The rehabilitation of Hazelwood mine for instance is expected to take a decade and is employing around 200 contractors".[83]

New Zealand's Climate Change Commissions draft advice report to the New Zealand government (full advice due on May 31, 2021)[84] when analysed from a restorative perspective approached the just transition in a different way again. Whilst saying that not making a just transition would be worse for jobs and the environment, the report acknowledged that if the New Zealand government was to make the just transition in line with its commitments on climate change, restoring all the jobs that would be lost would be, in the short term, challenging: "Some communities may see the closure of large businesses that provide significant employment for the community… To help understand the impact on employment, we commissioned a new model called the Distributional Impacts Microsimulation for Employment… This model cannot tell us about the aggregate effect on jobs in Aotearoa".[85]

The report also acknowledges the challenges that come with restoration of the natural environment, particularly New Zealand's plans to utilise afforestation: "we have heard through our engagement…concerns that the speed and potential extent of afforestation could have negative impacts on rural communities and provincial centres…reliant on the food and fibre industry for employment".[86] New Zealand's CCC argues that if the priority is to meet New Zealand's emissions and climate targets, in the short term, it may be difficult for the transition to be effective from a restorative perspective. However it does explain new industries shall emerge and the final report is due in one month wherein there could be updated advice in regard to a just transition's restorative benefits.

Finally other countries detail in a minor way the restorative benefits of a just transition in specifically affected regional areas (Germany being a prime example[87] but also ARC[88]), whereas others focus upon conserving jobs which already exist or essentially paying off workers who will have no jobs to step into post transition (Canada's pension bridging programme would be a typical example of such an approach[89]). Some, such as South Africa, make no overt references to the transition being restorative.[90] It is also important to note that restorative benefits where they do appear primarily manifest themselves in the form of new jobs replacing jobs which have disappeared, with environmental restoration being the second-order restoration priority. However as the ILO guidelines, the ACTUs arguments and Irelands NESC report show these are important aspects of a just transition and are not mutually exclusive to one another.

4.3.2 *Universal: Recognition and Cosmopolitanism*

4.3.2.1 Recognition Justice

A crucial component of what makes a just transition "just" is its recognition of the needs and rights of societal groups who are vulnerable within society generally as well as vulnerable specifically to the impact of a transition to a low-carbon/carbon-neutral society. In the context of JTCs, the main ways in which recognition can manifest is either through the composition of the JTC itself aka recognition manifesting itself through members of vulnerable, marginalised or disadvantaged people being represented by having seats on the commission or through specific recommendations or goals which the commissions advocate for or put forward in their reports and literature. As with all the other lenses through which the JTCs have so far been analysed, there are variations in the way in which recognition manifests in keeping with the varied nature of the countries and organisations.

New Zealand boasts what is perhaps the most unique composition of a commission and most unique recommendations in their report. This is due to the fact that 16.7% of New Zealand's population identify as being part of the Maori community.[91] The Maori are the indigenous people of mainland New Zealand, and whilst they form only just over 15% of the population in New Zealand, recognition of this community, their culture and traditions is very important, especially when making such a huge societal transition as New Zealand is seeking to do. As such, one of the CCC's seven commissioners, and deputy chairperson of the commission, is Ms Lisa Tumahai, who, the CCC's website explains, "is a leader within Ngāi Tahu…She also has established relationships with iwi and Māori and will help to ensure the broader Commission has a greater understanding of te ao Māori perspectives".[92]

Elsewhere, the CCC's most recent report features large sections of information which relates specifically to how a just transition in New Zealand would affect the Maori people (Section 5.6 of the draft advice for consultation[93]) and has translations throughout the report for members of the Maori community who speak te reo, the language of the Maori people. Whilst the JTC does a good job recognising the Maori people at both a commission- and recommendation-based level, none of the 7 commissioners are from a union/organised labour-based background which seems like oversight; however there are jobs and labour-related recommendations in the CCC's literature which to some extent could satisfy the recognition requirement of the crucially impacted group.

Interestingly, the ETA, as put forward by the ACTU, seems to suffer from the opposite problem to the CCC in that the ETA's creation has been put forward to the Australian Government by the Australian Council of Trade Unions, but it does not mention recognition of the Aboriginal Australian community. It does however say that an ETA should "Value the work of female-dominated industries... Better valuing the work of workers in female-dominated industries through higher wages and better conditions would reduce the impact of fossil-fuel plant closures on households and communities".[94] Recognition of Aboriginal Australians in the same manner as the CCC would be welcome though it is probably an unfair criticism given that the ETA is yet to even be established at the time of writing.

Canada is another country with an indigenous population, comprising 4.9% of the Canadian population,[95] and the Canadian JTC explicitly recognises this group, if not by having a member on the task force then through explicitly mentioning this group in their "Just Transition Supports For Communities: Key Messages" section of the report, saying that "First Nation and Métis communities may also be affected. Given Canada's commitment to reconciliation and advancing nation-to-nation relationships, it will be important to work in partnership with affected Indigenous communities to minimize any adverse impacts the phase-out may have".[96] Recognition of these groups, coupled with Canada's JTC's very strong labour/union/worker representation, both on the task force (7 out of the 11 task force members are from such backgrounds) and in their recommendations, makes Canada's JTC another very strong JTC from a recognition standpoint.

Having looked at the countries wherein there is a special category of indigenous community, by in large the remaining JTCs do a sufficient to strong job when it comes to recognition of workers, special category people and affected communities. The NESC's report to the Irish government states, "responding to climate change is an opportunity to engage in restorative justice and requires actively engaging vulnerable and underrepresented groups in terms of gender, ethnicity and socio-economic status while developing responses to climate change".[97]

Three council members of the NESC itself hail from the Irish Congress of Trade Unions,[98] as well as environmental, agricultural, business and academic councillors, suggesting commendable levels of recognition on the Irish JTC and in their report. Furthermore the Irish Just Transition Commissioner Kieran Mulvey was appointed following the closing down of peat-fired power stations in the Irish midlands, with a mandate for "guiding the development of the region in the aftermath of the ESB moving away from fossil-fuel generated power, a move which has sparked fears that areas of the midlands will be economically decimated in the fallout",[99] a move which displays a clear sense of recognition for the concerns of workers who may suffer from disruption during a just transition.

It could be argued that when compared to, for example, Canada, the JTCs for Scotland[100] and Germany[101] in terms of membership of their JTCs lean more towards technocratic, academic commissioners as opposed to those with a background in the labour and/or union movement. However this presence is not entirely absent and both JTCs display string recognition for this group in their reports and in the policies they advocate to their respective governments.

ARC displays a recognition angle through its focus upon the workers who are or may be suffering from economic strife due to the disappearance of fossil fuel industry jobs, when it comes to the work it undertakes promoting development in the Appalachian Regions,[102] whereas South Africa's JTC describes the composition of its JTC as being made up of representatives of labour and civil society, among other groups.[103]

However, whilst the degree of recognition on the JTCs themselves vary and will be subject to their own country or geographic region's unique demographics and challenges when it comes to a just transition, all recognise, at a minimum, workers and the ways in which they will be impacted and how these impacts must be mitigated, in their reports and advice. This is something common across all the countries and international organisations.

4.3.2.2 *Cosmopolitan Justice*

Climate change is a global problem which if left unaddressed could jeopardise the ability to maintain life on earth. The idea of a just transition is one of the primary solutions imagined for tackling climate change. Given the global scale of the problem, the solution, even when examining the problem and associated challenges when it comes to tackling it within the context of an individual country, cannot omit having a global, cosmopolitan perspective on the issue and the extent to which the actions of one country in tackling the issue impacts upon other countries across the world. As such, most of the countries and naturally the international institutions discuss the process of a just transition while making cosmopolitan references to the impact of their efforts beyond their own shores and borders.

Australia is one of the few countries which does not overtly discuss the cosmopolitan nature of a just transition, though the ACTU does make reference to Germany's coal phase-out approach as a model which Australia, if an ETA is ever established, should adapt.[104] In contrast to Australia's perspective, the Canadian task force's report for the Canadian government takes an overtly cosmopolitan perspective and links directly Canada's efforts to transition with the wider world: "Canada is not alone in experiencing the impacts of climate change or in taking action to reduce the GHG emissions...Over 180 countries are committed to doing their part to limit climate change...The Government of Canada has a window of opportunity...to not only adopt and implement our recommendations, but to demonstrate global leadership in "just transition" to a low-carbon economy".[105]

The final report of Germany's commission similarly displays a cosmopolitan bent and, similar to Canada, stresses the point that the countries "are not alone" in taking on the challenge of transitioning: "Germany is not alone in facing this challenge. A series of countries have already resolved to end coal-fired power generation and begun to take action... If Germany manages to successfully implement structural change processes here and to find the right balance between climate action, the creation of good jobs... then the energy transition and the associated ending of coal-fired power generation may provide an example for other countries".[106]

The NESC's just transition report for the Irish government also sets the stage for its recommendations by providing the global, cosmopolitan context in which the report was written thusly: "Countries across the globe are grappling with the shifts and challenges of climate change and environmental degradation and other transformative megatrends, such as urbanisation, population growth and globalisation".[107]

Part 1 of the New Zealand CCC's draft advice for the New Zealand Government is called "Our place in the world",[108] and similar to the

NESC's use of the global, cosmopolitan context of climate change to set the stage for its advice, the CCC discusses the problem of climate change in a global cosmopolitan fashion for three chapters before getting into New Zealand specific approaches for transitioning. Interestingly also, the New Zealand CCC, whilst acknowledging the cosmopolitan angle of the just transition, does not withhold criticism of other countries around the world for setting targets and failing to take necessary action to meet them: "Although countries worldwide have signed up to the Paris Agreement, current global efforts are not going far enough to bring about the emissions reductions needed. Reducing global emissions needs to be a collaborative effort".[109]

This focus on the cosmopolitan aspect beyond seeking to reassure a national government that it is "not alone" in the challenge it faces sets the New Zealand CCC apart from the other JTCs covered from a cosmopolitan perspective, in that it clearly believes the cosmopolitan aspect of transitioning is as important and worthy of proper focus as the procedural, distributive and other aspects and makes it clear where there are shortcomings in New Zealand or abroad, this JTC will not hesitate to point them out.

The South African Presidential Climate Change Coordinating Commission makes a passing reference to South Africa's emission reduction and adaption goals but does not specify if these are international or national, and by in large South Africa omits to focus on the global, cosmopolitan angle of the just transition.[110] ARC in the USA similarly omits focussing upon the cosmopolitan angle of their work though this is somewhat understandable given that ARC has a specific mandate to focus upon the needs of transitioning in a specific region of the USA and is not conceived of as being a body with a global outlook: "...address persistent economic disparities in the Region in comparison to the rest of the nation. PARC's mission was to draw up "a comprehensive program for the economic development of the Appalachian Region".[111]

Finally, the two international organisations of course display a focus upon the cosmopolitan, global angle of the just transition. The EU's just transition platform operates on the basis of helping to facilitate the region's transition to carbon neutrality in line with the EU's aim of "becoming the first climate-neutral bloc in the world by 2050".[112] The ILO early on in its document where it provides its transition guidelines states that despite countries employing different means for achieving a just transition, all countries share "a common global purpose".[113]

To conclude, the cosmopolitan element is present in almost all of the JTCs looked at, though for Australia and South Africa, this element is less overt. In contrast, New Zealand's CCC has a heavy cosmopolitan focus in the opening of its draft report and goes further than most other countries and even the international organisations in stressing the essentially cosmopolitan nature of the transition and in how it cannot be ignored if the transition globally is to succeed. Despite this variance, however, cosmopolitan elements can be seen in every JTC's approach to the challenge of transitioning within the context of its own borders, and whilst it may be minimal in some cases, it is positive to see that this aspect of the transition has not been overlooked.

4.3.3 *Space and Time*

Whilst the previous section analysed the extent to which all of the respective JTCs acknowledged the cosmopolitan, global nature of making just transitions to low-carbon or carbon-neutral economies, this final section shall look specifically at where in the world these initiatives are taking place and the deadlines that each respective JTC is working towards in the effort to make these transitions. In regard to space, speaking in general terms, whilst a number of these JTCs operate in the Southern Hemisphere, including New Zealand, South Africa and Australia, aside from South Africa, they all operate in what is referred to as the Global North, the Global North mostly being the Western, advanced, first-world nations where the majority of global wealth can be found. Speaking in general terms about time, almost all of the JTCs are working towards a transition which takes place during the course of this century though as we shall see there are variations when it comes to deadlines and the scopes of the transitions. In terms of time, a brief overview is as follows:

- **Australia**: ETA—Has not yet been established so presently no start or end date for the JTC's lifespan.
- **Canada**: Task Force on the Just Transition for Canadian Coal Power Workers and Communities—According to the Canadian government website, following the task force's provision of its two final reports on the transition in early 2019, the task force "completed its mandate",[114] and there has been no recent activity on its web page on the Canadian government website since that time (March 2019),

suggesting that the task force has concluded its work and is no longer operating.

- **The EU**: Just Transition Platform—Following its establishment in mid-2020, the just transition platform is still currently operating, having held its most recent meeting at the end of April 2021.[115] The EU Commission website does not indicate a fixed end or conclusion date for the platform's operation.
- **Germany**: Commission on Growth, Structural Change and Employment—Following the publication of its final report in early 2019, the commission, having completed its mandate, appears to no longer be operating.
- **International**: Guidelines—Following the tripartite meeting of experts in late 2015 to review, amend and adopt the draft guidelines, the guidelines were published, thereby concluding the task of the tripartite meeting of experts.[116]
- **Ireland**: Just Transition Review Group—NESC was asked to establish a Just Transition Review Group under its working group structures to advise the Climate Action Delivery Board. Following the completion of NESC's report in 2019, with the assistance of the Just Transition Review Group, one of the report's recommendations to the Irish government was for it to establish a Just Transition Review Group within NESC. In June 2020, Kieran Mulvey, the Irish Government's Just Transition Commissioner in his first report on the Just Transition in the midlands, explains: "NESC was also asked to establish a Just Transition Review Group under its working group structures to advise the Climate Action Delivery Board. It is intended that the Just Transition Review Group will be established for an initial period of 12 months and will provide strategic advice on the key elements of a just transition, drawing on national and international expertise."[117] Despite Mulvey being appointed just transition commissioner in late 2019,[118] the Just Transition Review Group within NESC, as of the time of writing, does not appear to have been established.
- **New Zealand**: Climate Change Commission—The commission was formed in late 2019 and still exists to this day, with a report due this year. According to the commission's website, "We take our whakapapa seriously and will understand and respect our origins in the coming years. This is just the beginning for the Commission and an important step on the long and urgent journey to secure a sustain-

able future for generations to come".[119] Its existence and operation shall be longer term and ongoing rather than, for example, the German and Canadian JTCs, with the CCC's website providing no fixed end date for its work.

- **Scotland**: Just Transition Commission—The commission delivered its final report in March 2021 at which point, according to the Scottish Government website, "the life of this Commission draws to a close".[120]
- **South Africa**: Presidential Climate Change Coordinating Commission—The commission was established in late 2020 and as of the moment has no fixed end date for the delivery of its work or mandate.
- **The USA**: Appalachian Regional Commission—The commission was established in the 1960s and is still in operation.

Australia, as previously stated, lags significantly behind the other countries of the Global North and now even South Africa in not having a JTC at all, rather there is just a proposal for a JTC provided by the ACTU. Given it is so far behind in this sense, the ACTU does not provide any date in its proposal for a transition, though it makes a vague reference to space when it says "Australia's energy transition is going to have the greatest impact on the workers, family members and communities that are employed in or are host to Australia's remaining 18 coal-fired power stations".[121] The specification of the 18 power stations gives some indication as to where the transition will take place, but no dates for when they ought to have transitioned by are provided.

Canada can be contrasted starkly against Australia as the Canadian JTC's final report makes reference to Canada's commitment "to the phase-out of traditional coal-fired electricity across the country by 2030"[122] before it goes on to provide a region-by-region breakdown of which regions and areas of Canada shall be most greatly impacted and therefore where the focus for the transition and the mitigation of any negative impacts must be: "Phasing out coal-fired electricity, however, will have direct and indirect impacts on thousands of workers, dozens of communities, and four provinces, including…Alberta, Saskatchewan, New Brunswick, and Nova Scotia…Nearly 50 communities with nearby coal mines or generating stations…3,000 to 3,900 workers at coal-fired generating stations and domestic thermal coal mines…Over a dozen generating stations, owned by six employers…Nine mines, owned by three employers".[123]

Similar to Canada when it comes to time, the German commission's report has a number of dates and deadlines that it is working towards when it comes to transition, explaining early on in its report that "Germany has set ambitious national climate action targets in order to largely achieve greenhouse gas neutrality by 2050".[124] More specifically, when it comes to the coal phase-out, the German commission recommends "…the end of 2038 as the end date for coal-fired power generation. If the requirements for energy, employment and the economy are met, the date may be brought forward to no earlier than 2035 in negotiations with the operators. This question will be reviewed in 2032 to decide whether this is possible",[125] providing a clear timeline for the review of progress and eventual phase-out of coal. Regarding time, like Canada, the German report also features a detailed breakdown of the regions of the country which shall be most strongly affected and therefore must be strongly focussed upon during the transition.

In the Global South, the South African presidential commission does not have any specifics when it comes to dates for transitioning nor does it provide specifics in terms of the areas where the transition will focus;[126] however these could possibly become clearer as the work of the commission gets underway as it has only fairly recently been established.

The ARC and the EU differ from the other global northern nations and South Africa as well. In terms of space, ARC's work focuses on the region and states of America known as the Appalachian Region, an area that cuts across 13 states in America, setting it apart from all the other JTCs bar the EU and the ILO in not being an entirely national transition commission.[127] Unlike the other commissions on this list, ARC also does not set a date for when the region it focuses on should transition away from coal. The EU differs from ARC, the rest of the global North and South Africa in that whilst the EU comprises of 27 member states, its JTC is primarily focussed on specific regions of specific nations: "These territories will need to restructure and/or diversify their economy, maintain social cohesion, and (re)train the affected workers and youth to prepare them for future jobs".[128] As well as this, the EU has committed itself to "becoming the first climate-neutral bloc in the world by 2050",[129] denoting the date and broad area for where and when the transition is due to happen.

Overall, there are significant differences when it comes to the geographical areas each respective country or organisations JTC covers and the time by which transitions are aimed to be completed by. On the other hand though, with the exception of South Africa, ARC in the USA and Australia, the majority of the remaining JTCs have a specific geographic region or area of focus as well as deadlines and time commitments by which the aim is to see that the transition is completed. Germany and Canada are among the most specific when it comes to space and time and provide good examples as to how specifics, when it comes to these factors, are important for realising a just transition rather than simply invoking the concept without concrete data and analysis behind it to make it happen.

4.4 Conclusion for JTCs

What becomes clear as one analyses the JTCs from the different perspectives is that none of them are completely perfect on all measures, and in some instances, the JTCs are quite difficult to compare due to the significant differences that can be present when it comes to the scope and ambition of different JTCs. However, there are broad commonalities, and along these lines of commonality, one can attempt to assess which JTCs are effective (or likely to be effective) and which shall struggle.

New Zealand stand outs as the JTC which is likely to be the most effective. This is because not only does it perform strongly when subject to the various analysis but also on account of the fact that the New Zealand JTC is a body which still exists and will continue during its life span to scrutinise and advise the New Zealand government when it comes to making a just transition. It has also, compared to the other JTCs, displayed that it is not afraid to criticise the New Zealand government or other governments across the world, when it comes to actions failing to match rhetoric or commitments. This approach to its role, whilst arguably uncompromising, displays a sense of seriousness the New Zealand JTC feels about the just transition that is matched by its procedural coherency, distributive strength, its honest analysis when it comes to the difficulties of a restorative transition in New Zealand, its excellent recognition levels as well as its strong emphasis on the cosmopolitan nature of the just transition. Also, it is guided by dates and targets which its draft consultation suggests it will rigidly attempt to hold the New Zealand government to. All of this makes it more likely that the New Zealand JTC shall prove effective at guiding its country's government through a just transition.

On the opposite end of the spectrum is Australia whose JTC appears rather limited, primarily due to the fact that the Australian JTC presently only exists in a conceptual form argued for by the ACTU. As such at this point, Australia is lagging behind even South Africa, the only country whose JTC is from the Global South and expect much more to develop in South Africa as they are very ambitious with their just transition plans.

The remaining JTCs fit somewhere in between Australia and New Zealand in terms of how likely they are to be. Germany and Scotland are both strong though it could be argued that they could be improved from a recognition perspective. Ireland and Canada both have strong suits but lack strongly in other areas, particularly from a distributive analysis in Ireland's case and in Canada's case, due to a lack of focus upon creating good sustainable jobs when it comes to the recommendations of its JTC.

Overall, each JTC has something to recommend it, and analysis from all the different perspectives corroborates the assertion of the ILO when it states that when it comes to JTCs, "There is no 'one-size-fits-all'".

Notes

1. Davey, M., 2016. *Hazelwood coal power station to close with loss of up to 1,000 jobs.* [online] theguardian.com. Available at: https://www.theguardian.com/australia-news/2016/nov/03/hazelwood-coal-power-station-to-close-with-loss-of-800-jobs (last accessed 1 September 2021).
2. Premier of Victoria, 2016. *Labor Government To Support Hazelwood Workers | Premier of Victoria.* [online] Premier.vic.gov.au. Available at: https://www.premier.vic.gov.au/labor-government-support-hazelwood-workers (last accessed 1 September 2021).
3. Ministers for the Department of Industry, Science, Energy and Resources, 2016. *Government to support Hazelwood workers.* [online] minister.industry.gov.au. Available at: https://www.minister.industry.gov.au/ministers/hunt/media-releases/government-support-hazelwood-workers (last accessed 1 September 2021).
4. ACTU, 2021. *The Need for a Just Transition.* [online] Actu.org.au. Available at: https://www.actu.org.au/our-work/climate-change/the-need-for-a-just-transition (last accessed 1 September 2021).
5. Environment and Communications References Committee, 2017. *Retirement of coal fired power stations.* [ebook] Available at: https://www.aph.gov.au/. (last accessed 1 September 2021).

6. Environment and Communications References Committee, 2017. *Retirement of coal fired power stations.* [ebook] https://www.aph.gov.au/. Available at: http://file:///C:/Users/O'Boyle%20family/Downloads/report.pdf (last accessed 1 September 2021).
7. Hériault, C., 2018. *Canada's coal power phase-out reaches another milestone—Canada.ca.* [online] Canada.ca. Available at: https://www.canada.ca/en/environment-climate-change/news/2018/12/canadas-coal-power-phase-out-reaches-another-milestone.html (last accessed 1 September 2021).
8. Government of Canada, 2018. *Information archivée dans le Web | Information Archived on the Web.* [online] Publications.gc.ca. Available at: http://publications.gc.ca/collections/collection_2019/eccc/En4-361-2019-eng.pdf (last accessed 1 September 2021).
9. Government of Canada, 2018. *Information archivée dans le Web | Information Archived on the Web.* [online] Publications.gc.ca. Available at: http://publications.gc.ca/collections/collection_2019/eccc/En4-361-2019-eng.pdf (last accessed 1 September 2021).
10. European Commission, 2017. *No region left behind: launch of the Platform for Coal Regions in Transition.* [online] ec.europa.eu. Available at: https://ec.europa.eu/commission/presscorner/detail/en/IP_17_5165 (last accessed 1 September 2021).
11. European Commission, 2017. *No region left behind: launch of the Platform for Coal Regions in Transition.* [online] ec.europa.eu. Available at: https://ec.europa.eu/commission/presscorner/detail/en/IP_17_5165 (last accessed 1 September 2021).
12. Colucci, A., 2018. *Coal Regions in Transition Platform.* [online] Ec.europa.eu. Available at: https://ec.europa.eu/energy/sites/ener/files/documents/1_eu_coal_regions_in_transition.pdf (last accessed 1 September 2021).
13. European Commission, 2020. *Just Transition Platform.* [online] ec.europa.eu. Available at: https://ec.europa.eu/info/strategy/priorities-2019-2024/european-green-deal/actions-being-taken-eu/just-transition-mechanism/just-transition-platform_en#about-the-just-transition-platform (last accessed 1 September 2021).
14. European Commission, 2020. *Just Transition Platform.* [online] ec.europa.eu. Available at: https://ec.europa.eu/info/strategy/priorities-2019-2024/european-green-deal/actions-being-taken-eu/just-transition-mechanism/just-transition-platform_en#about-the-just-transition-platform (last accessed 1 September 2021).
15. Federal Ministry for the Environment, Nature Conservation and Nuclear Safety, 2016. *Climate Action Plan 2050—Germany's long-term low greenhouse gas emission development strategy.* [online] www.bmu.de. Available

at: https://www.bmu.de/en/topics/climate-energy/climate/national-climate-policy/greenhouse-gas-neutral-germany-2050/ (last accessed 1 September 2021).

16. Federal Ministry for the Environment, Nature Conservation and Nuclear Safety, 2016. *Climate Action Plan 2050—Germany's long-term low greenhouse gas emission development strategy*. [online] www.bmu.de. Available at: https://www.bmu.de/en/topics/climate-energy/climate/national-climate-policy/greenhouse-gas-neutral-germany-2050/ (last accessed 1 September 2021).
17. German Government, 2018. *Launch of Commission on Growth, Structural Change and Employment*. [online] Bmu.de. Available at: https://www.bmu.de/fileadmin/Daten_BMU/Download_PDF/Klimaschutz/einsetzungsbeschluss_kohlekommission_en_bf.pdf (last accessed 1 September 2021).
18. Wehrmann, B., 2018. *Germany's coal exit commission*. [online] https://www.cleanenergywire.org/. Available at: https://www.cleanenergywire.org/factsheets/germanys-coal-exit-commission (last accessed 1 September 2021).
19. Commission on Growth, Structural Change and Employment, 2019. *Final Report*. [online] Bmwi.de. Available at: https://www.bmwi.de/Redaktion/EN/Publikationen/commission-on-growth-structural-change-and-employment.pdf?__blob=publicationFile&v=3 (last accessed 1 September 2021).
20. United Nations, 2015. *Paris Agreement*. [online] Unfccc.int. Available at: https://unfccc.int/sites/default/files/english_paris_agreement.pdf (last accessed 1 September 2021).
21. International Labour Organisation, 2015. *Guidelines for a just transition towards environmentally sustainable economies and societies for all*. [online] Ilo.org. Available at: https://www.ilo.org/wcmsp5/groups/public/%2D%2D-ed_emp/%2D%2D-emp_ent/documents/publication/wcms_432859.pdf (last accessed 1 September 2021).
22. International Labour Organisation, 2015. *Guidelines for a just transition towards environmentally sustainable economies and societies for all*. [online] Ilo.org. Available at: https://www.ilo.org/wcmsp5/groups/public/%2D%2D-ed_emp/%2D%2D-emp_ent/documents/publication/wcms_432859.pdf (last accessed 1 September 2021).
23. International Labour Organisation, 2015. *Guidelines for a just transition towards environmentally sustainable economies and societies for all*. [online] Ilo.org. Available at: https://www.ilo.org/wcmsp5/groups/public/%2D%2D-ed_emp/%2D%2D-emp_ent/documents/publication/wcms_432859.pdf (last accessed 1 September 2021).

24. United Nations, 2021. *UN Climate Change Conference (COP26) at the SEC—Glasgow 2021*. [online] https://ukcop26.org/. Available at: https://ukcop26.org/ (last accessed 1 September 2021).
25. CLIMATE ACTION PLAN, 2019. *CLIMATE ACTION PLAN 2019 To Tackle Climate Breakdown*. [online] Assets.gov.ie. Available at: https://assets.gov.ie/25419/c97cdecddf8c49ab976e773d4e11e515.pdf (last accessed 1 September 2021).
26. CLIMATE ACTION PLAN, 2019. *CLIMATE ACTION PLAN 2019 To Tackle Climate Breakdown*. [online] Assets.gov.ie. Available at: https://assets.gov.ie/25419/c97cdecddf8c49ab976e773d4e11e515.pdf (last accessed 1 September 2021).
27. NESC, 2020. *Addressing Employment Vulnerability as Part of a Just Transition in Ireland*. [online] Files.nesc.ie. Available at: http://files.nesc.ie/nesc_reports/en/149_Transition.pdf (last accessed 1 September 2021).
28. NESC, 2020. *Addressing Employment Vulnerability as Part of a Just Transition in Ireland*. [online] Files.nesc.ie. Available at: http://files.nesc.ie/nesc_reports/en/149_Transition.pdf (last accessed 1 September 2021).
29. NESC, 2020. *Addressing Employment Vulnerability as Part of a Just Transition in Ireland*. [online] Files.nesc.ie. Available at: http://files.nesc.ie/nesc_reports/en/149_Transition.pdf (last accessed 1 September 2021).
30. ICCC, 2018. *Our work | Interim Climate Change Committee*. [online] Iccc.mfe.govt.nz. Available at: https://www.iccc.mfe.govt.nz/what-we-do/our-work/ (last accessed 1 September 2021).
31. ICCC, 2018. *Our work | Interim Climate Change Committee*. [online] Iccc.mfe.govt.nz. Available at: https://www.iccc.mfe.govt.nz/what-we-do/our-work/ (last accessed 1 September 2021).
32. Climate Change Commission, 2019. *He Pou a Rangi—Climate Change Commission*. [online] Climate Change Commission. Available at: https://www.climatecommission.govt.nz/ (last accessed 1 September 2021).
33. Climate Change Commission, 2019. *He Pou a Rangi—Climate Change Commission*. [online] Climate Change Commission. Available at: https://www.climatecommission.govt.nz/ (last accessed 1 September 2021).
34. Climate Change Commission, 2019. *He Pou a Rangi—Climate Change Commission*. [online] Climate Change Commission. Available at: https://www.climatecommission.govt.nz/ (last accessed 1 September 2021).

35. Climate Change Commission, 2021. *Executive Summary: work must start now*. [online] Climate Change Commission. Available at: https://ccc-production-media.s3.ap-southeast-2.amazonaws.com/public/Executive-Summary-advice-report-v3.pdf (last accessed 1 September 2021).
36. Climate Change Commission, 2021. *Executive Summary: work must start now*. [online] Climate Change Commission. Available at: https://ccc-production-media.s3.ap-southeast-2.amazonaws.com/public/Executive-Summary-advice-report-v3.pdf (last accessed 1 September 2021).
37. Scottish Government, 2017. *Just Transition Commission*. [online] Gov.scot. Available at: https://www.gov.scot/groups/just-transition-commission/ (last accessed 1 September 2021).
38. Scottish Government, 2017. *Just Transition Commission*. [online] Gov.scot. Available at: https://www.gov.scot/groups/just-transition-commission/ (last accessed 1 September 2021).
39. Scottish Government, 2017. *Just Transition Commission*. [online] Gov.scot. Available at: https://www.gov.scot/groups/just-transition-commission/ (last accessed 1 September 2021).
40. Scottish Government, 2017. *Just Transition Commission*. [online] Gov.scot. Available at: https://www.gov.scot/groups/just-transition-commission/ (last accessed 1 September 2021).
41. Just Transition Commission, 2021. *Just Transition Commission A national mission for a fairer, greener Scotland*. [online] Gov.scot. Available at: https://www.gov.scot/binaries/content/documents/govscot/publications/advice-and-guidance/2021/03/transition-commission-national-mission-fairer-greener-scotland/documents/transition-commission-national-mission-fairer-greener-scotland/transition-commission-national-mission-fairer-greener-scotland/govscot%3Adocument/transition-commission-national-mission-fairer-greener-scotland.pdf (last accessed 1 September 2021).
42. Just Transition Commission, 2021. *Just Transition Commission A national mission for a fairer, greener Scotland*. [online] Gov.scot. Available at: https://www.gov.scot/binaries/content/documents/govscot/publications/advice-and-guidance/2021/03/transition-commission-national-mission-fairer-greener-scotland/documents/transition-commission-national-mission-fairer-greener-scotland/transition-commission-national-mission-fairer-greener-scotland/govscot%3Adocument/transition-commission-national-mission-fairer-greener-scotland.pdf (last accessed 1 September 2021).
43. National Planning Commission, 2010. *Home*. [online] Nationalplanningcommission.org.za. Available at: https://www.nation-

alplanningcommission.org.za/Pages/default (last accessed 1 September 2021).
44. The Presidency Republic of South Africa, 2020. *Presidential Climate Change Coordinating Commission appointed.* [online] http://www.thepresidency.gov.za/press-statements/presidential-climate-change-coordinating-commission-appointed. Available at: http://www.thepresidency.gov.za/press-statements/presidential-climate-change-coordinating-commission-appointed (last accessed 1 September 2021).
45. The Presidency Republic of South Africa, 2020. *Presidential Climate Change Coordinating Commission appointed.* [online] http://www.thepresidency.gov.za/press-statements/presidential-climate-change-coordinating-commission-appointed. Available at: http://www.thepresidency.gov.za/press-statements/presidential-climate-change-coordinating-commission-appointed (last accessed 1 September 2021).
46. The Presidency Republic of South Africa, 2020. *Presidential Climate Change Coordinating Commission appointed.* [online] http://www.thepresidency.gov.za/press-statements/presidential-climate-change-coordinating-commission-appointed. Available at: http://www.thepresidency.gov.za/press-statements/presidential-climate-change-coordinating-commission-appointed (last accessed 1 September 2021).
47. Appalachian Regional Commission, n.d. *About the Appalachian Regional Commission—Appalachian Regional Commission.* [online] Appalachian Regional Commission. Available at: https://www.arc.gov/about/index.asp (last accessed 1 September 2021).
48. Appalachian Regional Commission, 2021. *About the Appalachian Region—Appalachian Regional Commission.* [online] https://www.arc.gov/about-the-appalachian-region/. Available at: https://www.arc.gov/about-the-appalachian-region/ (last accessed 1 September 2021).
49. Jones, S., 2017. *Appalachia is trying to diversify its economy. Trump isn't helping.* [online] newrepublic.com. Available at: https://newrepublic.com/article/141679/appalachia-trying-diversify-economy-trump-isnt-helping (last accessed 1 September 2021).
50. Jones, S., 2017. *Appalachia is trying to diversify its economy. Trump isn't helping.* [online] newrepublic.com. Available at: https://newrepublic.com/article/141679/appalachia-trying-diversify-economy-trump-isnt-helping (last accessed 1 September 2021).
51. Godfrey, E., 2017. *Trump's Proposal to Scrap the Agency Devoted to Developing Appalachia.* [online] https://www.theatlantic.com/. Available at: https://www.theatlantic.com/politics/archive/2017/03/why-the-appalachian-regional-commission-matters/519876/ (last accessed 1 September 2021).

52. Scottish Government, 2020. *Just Transition Commission Interim Report—gov.scot.* [online] Gov.scot. Available at: https://www.gov.scot/publications/transition-commission-interim-report/ (last accessed 1 September 2021).
53. ICCC, 2019. *Final Report to the Minister for Climate Change.* [online] Iccc.mfe.govt.nz. Available at: https://www.iccc.mfe.govt.nz/assets/70bfa26241/20-December-2020-Report-to-Minister-Shaw_web.pdf (last accessed 1 September 2021).
54. German Commission on Growth, Structural Change and Employment, 2018. *Role and Perspectives for German Coal Regions.* [online] Ec.europa.eu. Available at: https://ec.europa.eu/energy/sites/ener/files/documents/5_german_commission_on_growth_structural_change_and_employment-_role_and_perspectives_for_german_coal_regions_dr._philipp_steinberg_director_gener.pdf (last accessed 1 September 2021).
55. Just Transition Commission, 2021. *Just Transition Commission A national mission for a fairer, greener Scotland.* [online] Gov.scot. Available at: https://www.gov.scot/binaries/content/documents/govscot/publications/advice-and-guidance/2021/03/transition-commission-national-mission-fairer-greener-scotland/documents/transition-commission-national-mission-fairer-greener-scotland/transition-commission-national-mission-fairer-greener-scotland/govscot%3Adocument/transition-commission-national-mission-fairer-greener-scotland.pdf (last accessed 1 September 2021).
56. ICCC, 2019. *Final Report to the Minister for Climate Change.* [online] Iccc.mfe.govt.nz. Available at: https://www.iccc.mfe.govt.nz/assets/70bfa26241/20-December-2020-Report-to-Minister-Shaw_web.pdf (last accessed 1 September 2021).
57. Commission on Growth, Structural Change and Employment, 2019. *Final Report.* [online] Bmwi.de. Available at: https://www.bmwi.de/Redaktion/EN/Publikationen/commission-on-growth-structural-change-and-employment.pdf?__blob=publicationFile&v=3 (last accessed 1 September 2021).
58. Government of Canada, 2018. *Information archivée dans le Web | Information Archived on the Web.* [online] Publications.gc.ca. Available at: http://publications.gc.ca/collections/collection_2019/eccc/En4-361-2019-eng.pdf (last accessed 1 September 2021).
59. NESC, 2020. *Addressing Employment Vulnerability as Part of a Just Transition in Ireland.* [online] Files.nesc.ie. Available at: http://files.nesc.ie/nesc_reports/en/149_Transition.pdf (last accessed 1 September 2021).
60. Brennan, C., 2021. *Kieran Mulvey appointed Just Transition Commissioner for closure of peat-fired power stations.* [online] irishexaminer.com.

Available at: https://www.irishexaminer.com/news/arid-30962706.html (last accessed 1 September 2021).

61. Mulvey, K., 2020. *Just Transition Progress Report.* [ebook] Just Transition Commissioner. Available at: http://file:///C:/Users/O'Boyle%20family/Downloads/76825_2aedd89e-484c-4dea-a341-9d9036b46fd7.pdf (last accessed 1 September 2021).
62. Scottish Government, 2021. *Just Transition Commission: A National Mission for a fairer, greener Scotland—gov.scot.* [online] Gov.scot. Available at: https://www.gov.scot/publications/transition-commission-national-mission-fairer-greener-scotland/pages/6/ (last accessed 1 September 2021).
63. Environment and Communications References Committee, 2017. *Retirement of coal fired power stations.* [ebook] https://www.aph.gov.au/. Available at: http://file:///C:/Users/O'Boyle%20family/Downloads/report.pdf (last accessed 1 September 2021).
64. The Presidency Republic of South Africa, 2020. *Presidential Climate Change Coordinating Commission appointed.* [online] http://www.the-presidency.gov.za/press-statements/presidential-climate-change-coordinating-commission-appointed. Available at: http://www.thepresidency.gov.za/press-statements/presidential-climate-change-coordinating-commission-appointed (last accessed 1 September 2021).
65. Appalachian Regional Commission, 2021. *About the Appalachian Region—Appalachian Regional Commission.* [online] https://www.arc.gov/about-the-appalachian-region/. Available at: https://www.arc.gov/about-the-appalachian-region/ (last accessed 1 September 2021).
66. European Commission, 2020. *Just Transition Platform.* [online] ec.europa.eu. Available at: https://ec.europa.eu/info/strategy/priorities-2019-2024/european-green-deal/actions-being-taken-eu/just-transition-mechanism/just-transition-platform_en#about-the-just-transition-platform (last accessed 1 September 2021).
67. International Labour Organisation, 2015. *Guidelines for a just transition towards environmentally sustainable economies and societies for all.* [online] Ilo.org. Available at: https://www.ilo.org/wcmsp5/groups/public/%2D%2D-ed_emp/%2D%2D-emp_ent/documents/publication/wcms_432859.pdf (last accessed 1 September 2021).
68. Government of Canada, 2018. *Information archivée dans le Web | Information Archived on the Web.* [online] Publications.gc.ca. Available at: http://publications.gc.ca/collections/collection_2019/eccc/En4-361-2019-eng.pdf (last accessed 1 September 2021).
69. Commission on Growth, Structural Change and Employment, 2019. *Final Report.* [online] Bmwi.de. Available at: https://www.bmwi.de/Redaktion/EN/Publikationen/commission-on-growth-structural-

change-and-employment.pdf?__blob=publicationFile&v=3 (last accessed 1 September 2021).

70. Commission on Growth, Structural Change and Employment, 2019. *Final Report.* [online] Bmwi.de. Available at: https://www.bmwi.de/Redaktion/EN/Publikationen/commission-on-growth-structural-change-and-employment.pdf?__blob=publicationFile&v=3 (last accessed 1 September 2021).
71. Commission on Growth, Structural Change and Employment, 2019. *Final Report.* [online] Bmwi.de. Available at: https://www.bmwi.de/Redaktion/EN/Publikationen/commission-on-growth-structural-change-and-employment.pdf?__blob=publicationFile&v=3 (last accessed 1 September 2021).
72. NESC, 2020. *Addressing Employment Vulnerability as Part of a Just Transition in Ireland.* [online] Files.nesc.ie. Available at: http://files.nesc.ie/nesc_reports/en/149_Transition.pdf (last accessed 1 September 2021).
73. NESC, 2020. *Addressing Employment Vulnerability as Part of a Just Transition in Ireland.* [online] Files.nesc.ie. Available at: http://files.nesc.ie/nesc_reports/en/149_Transition.pdf (last accessed 1 September 2021).
74. ARC, n.d. *ARC's History and Work in Appalachia—Appalachian Regional Commission.* [online] arc.gov. Available at: https://www.arc.gov/arcs-history-and-work-in-appalachia/ (last accessed 1 September 2021).
75. The Presidency Republic of South Africa, 2020. *Presidential Climate Change Coordinating Commission appointed.* [online] http://www.thepresidency.gov.za/press-statements/presidential-climate-change-coordinating-commission-appointed. Available at: http://www.thepresidency.gov.za/press-statements/presidential-climate-change-coordinating-commission-appointed (last accessed 1 September 2021).
76. International Labour Organisation, 2015. *Guidelines for a just transition towards environmentally sustainable economies and societies for all.* [online] Ilo.org. Available at: https://www.ilo.org/wcmsp5/groups/public/%2D%2D-ed_emp/%2D%2D-emp_ent/documents/publication/wcms_432859.pdf (last accessed 1 September 2021).
77. International Labour Organisation, 2015. *Guidelines for a just transition towards environmentally sustainable economies and societies for all.* [online] Ilo.org. Available at: https://www.ilo.org/wcmsp5/groups/public/%2D%2D-ed_emp/%2D%2D-emp_ent/documents/publication/wcms_432859.pdf (last accessed 1 September 2021).
78. International Labour Organisation, 2015. *Guidelines for a just transition towards environmentally sustainable economies and societies for all.* [online]

Ilo.org. Available at: https://www.ilo.org/wcmsp5/groups/public/%2D%2D-ed_emp/%2D%2D-emp_ent/documents/publication/wcms_432859.pdf (last accessed 1 September 2021).
79. International Labour Organisation, 2015. *Guidelines for a just transition towards environmentally sustainable economies and societies for all.* [online] Ilo.org. Available at: https://www.ilo.org/wcmsp5/groups/public/%2D%2D-ed_emp/%2D%2D-emp_ent/documents/publication/wcms_432859.pdf (last accessed 1 September 2021).
80. NESC, 2020. *Addressing Employment Vulnerability as Part of a Just Transition in Ireland.* [online] Files.nesc.ie. Available at: http://files.nesc.ie/nesc_reports/en/149_Transition.pdf (last accessed 1 September 2021).
81. NESC, 2020. *Addressing Employment Vulnerability as Part of a Just Transition in Ireland.* [online] Files.nesc.ie. Available at: http://files.nesc.ie/nesc_reports/en/149_Transition.pdf (last accessed 1 September 2021).
82. ACTU, 2021. *The Need for a Just Transition.* [online] Actu.org.au. Available at: https://www.actu.org.au/our-work/climate-change/the-need-for-a-just-transition (last accessed 1 September 2021).
83. ACTU, 2021. *The Need for a Just Transition.* [online] Actu.org.au. Available at: https://www.actu.org.au/our-work/climate-change/the-need-for-a-just-transition (last accessed 1 September 2021).
84. Climate Change Commission, 2021. *He Pou a Rangi—Climate Change Commission.* [online] Climatecommission.govt.nz. Available at: https://www.climatecommission.govt.nz/get-involved/our-advice-and-evidence/ (last accessed 1 September 2021).
85. Climate Change Commission, 2021. *31 January 2021 Draft Advice for Consultation.* [online] Ccc-production-media.s3.ap-southeast-2.amazonaws.com. Available at: https://ccc-production-media.s3.ap-southeast-2.amazonaws.com/public/evidence/advice-report-DRAFT-1ST-FEB/ADVICE/CCC-ADVICE-TO-GOVT-31-JAN-2021-pdf.pdf (last accessed 1 September 2021).
86. Climate Change Commission, 2021. *31 January 2021 Draft Advice for Consultation.* [online] Ccc-production-media.s3.ap-southeast-2.amazonaws.com. Available at: https://ccc-production-media.s3.ap-southeast-2.amazonaws.com/public/evidence/advice-report-DRAFT-1ST-FEB/ADVICE/CCC-ADVICE-TO-GOVT-31-JAN-2021-pdf.pdf (last accessed 1 September 2021).
87. Commission on Growth, Structural Change and Employment, 2019. *Final Report.* [online] Bmwi.de. Available at: https://www.bmwi.de/Redaktion/EN/Publikationen/commission-on-growth-structural-

change-and-employment.pdf?__blob=publicationFile&v=3 (last accessed 1 September 2021).

88. ARC, n.d. *ARC's History and Work in Appalachia—Appalachian Regional Commission.* [online] arc.gov. Available at: https://www.arc.gov/arcs-history-and-work-in-appalachia/ (last accessed 1 September 2021).
89. Government of Canada, 2018. *Information archivée dans le Web | Information Archived on the Web.* [online] Publications.gc.ca. Available at: http://publications.gc.ca/collections/collection_2019/eccc/En4-361-2019-eng.pdf (last accessed 1 September 2021).
90. The Presidency Republic of South Africa, 2020. *Presidential Climate Change Coordinating Commission appointed.* [online] http://www.thepresidency.gov.za/press-statements/presidential-climate-change-coordinating-commission-appointed. Available at: http://www.thepresidency.gov.za/press-statements/presidential-climate-change-coordinating-commission-appointed (last accessed 1 September 2021).
91. Stats NZ, 2020. *Māori population estimates: At 30 June 2020 | Stats NZ.* [online] Stats.govt.nz. Available at: https://www.stats.govt.nz/information-releases/maori-population-estimates-at-30-june-2020#:~:text=New%20Zealand's%20estimated%20M%C4%81ori%20ethnic,males%20and%20426%2C800%20M%C4%81ori%20females. (last accessed 1 September 2021).
92. Climate Change Commission, 2021. *He Pou a Rangi—Climate Change Commission.* [online] Climatecommission.govt.nz. Available at: https://www.climatecommission.govt.nz/who-we-are/our-people-2/ (last accessed 1 September 2021).
93. Climate Change Commission, 2021. *He Pou a Rangi—Climate Change Commission.* [online] Climatecommission.govt.nz. Available at: https://www.climatecommission.govt.nz/get-involved/our-advice-and-evidence/ (last accessed 1 September 2021).
94. ACTU, 2021. *The Need for a Just Transition.* [online] Actu.org.au. Available at: https://www.actu.org.au/our-work/climate-change/the-need-for-a-just-transition (last accessed 1 September 2021).
95. OECD Library, 2021. *Profile of Indigenous Canada: Trends and data needs | Linking Indigenous Communities with Regional Development in Canada | OECD iLibrary.* [online] Oecd-ilibrary.org. Available at: https://www.oecd-ilibrary.org/sites/e6cc8722-en/index.html?itemId=/content/component/e6cc8722-en#:~:text=Canada%20has%20a%20diverse%20and,the%20Inuit%20population%20(4%25). (last accessed 1 September 2021).
96. Government of Canada, 2018. *Information archivée dans le Web | Information Archived on the Web.* [online] Publications.gc.ca. Available

at: http://publications.gc.ca/collections/collection_2019/eccc/En4-361-2019-eng.pdf (last accessed 1 September 2021).
97. NESC, 2020. *Addressing Employment Vulnerability as Part of a Just Transition in Ireland.* [online] Files.nesc.ie. Available at: http://files.nesc.ie/nesc_reports/en/149_Transition.pdf (last accessed 1 September 2021).
98. NESC, 2021. *The Council—National Economic & Social Council.* [online] National Economic & Social Council. Available at: https://www.nesc.ie/our-organisation/the-council/ (last accessed 1 September 2021).
99. Brennan, C., 2021. *Kieran Mulvey appointed Just Transition Commissioner for closure of peat-fired power stations.* [online] irishexaminer.com. Available at: https://www.irishexaminer.com/news/arid-30962706.html (last accessed 1 September 2021).
100. Just Transition Commission, 2021. *Just Transition Commission: member profiles—gov.scot.* [online] Gov.scot. Available at: https://www.gov.scot/publications/just-transition-commission-member-profiles/ (last accessed 1 September 2021).
101. Federal Ministry for Economic Affairs and Energy, 2021. *Commission on Growth, Structural Change and Employment Final Report.* [online] Bmwi.de. Available at: https://www.bmwi.de/Redaktion/EN/Publikationen/commission-on-growth-structural-change-and-employment.pdf?__blob=publicationFile&v=3 (last accessed 1 September 2021).
102. Appalachian Regional Commission, 2021. *About the Appalachian Region—Appalachian Regional Commission.* [online] https://www.arc.gov/about-the-appalachian-region/. Available at: https://www.arc.gov/about-the-appalachian-region/ (last accessed 1 September 2021).
103. The Presidency Republic of South Africa, 2020. *Presidential Climate Change Coordinating Commission appointed.* [online] http://www.thepresidency.gov.za/press-statements/presidential-climate-change-coordinating-commission-appointed. Available at: http://www.thepresidency.gov.za/press-statements/presidential-climate-change-coordinating-commission-appointed (last accessed 1 September 2021).
104. ACTU, 2021. *The Need for a Just Transition.* [online] Actu.org.au. Available at: https://www.actu.org.au/our-work/climate-change/the-need-for-a-just-transition (last accessed 1 September 2021).
105. Government of Canada, 2018. *Information archivée dans le Web | Information Archived on the Web.* [online] Publications.gc.ca. Available at: http://publications.gc.ca/collections/collection_2019/eccc/En4-361-2019-eng.pdf (last accessed 1 September 2021).

106. Commission on Growth, Structural Change and Employment, 2019. *Final Report.* [online] Bmwi.de. Available at: https://www.bmwi.de/Redaktion/EN/Publikationen/commission-on-growth-structural-change-and-employment.pdf?__blob=publicationFile&v=3 (last accessed 1 September 2021).
107. NESC, 2020. *Addressing Employment Vulnerability as Part of a Just Transition in Ireland.* [online] Files.nesc.ie. Available at: http://files.nesc.ie/nesc_reports/en/149_Transition.pdf [Accessed 22 April 2021].
108. Climate Change Commission, 2021. *He Pou a Rangi—Climate Change Commission.* [online] Climatecommission.govt.nz. Available at: https://www.climatecommission.govt.nz/get-involved/our-advice-and-evidence/ (last accessed 1 September 2021).
109. The Climate Change Commission, 2021. *What are other countries doing?.* [online] Ccc-production-media.s3.ap-southeast-2.amazonaws.com. Available at: https://ccc-production-media.s3.ap-southeast-2.amazonaws.com/public/evidence/advice-report-DRAFT-1ST-FEB/Evidence-CH-02-what-are-other-countries-doing-21-Jan-2021.pdf [(last accessed 1 September 2021).
110. The Presidency Republic of South Africa, 2020. *Presidential Climate Change Coordinating Commission appointed.* [online] http://www.thepresidency.gov.za/press-statements/presidential-climate-change-coordinating-commission-appointed. Available at: http://www.thepresidency.gov.za/press-statements/presidential-climate-change-coordinating-commission-appointed (last accessed 1 September 2021).
111. ARC, 2021. *ARC's History and Work in Appalachia—Appalachian Regional Commission.* [online] Appalachian Regional Commission. Available at: https://www.arc.gov/arcs-history-and-work-in-appalachia/ (last accessed 1 September 2021).
112. The European Commission, 2021. *Press corner.* [online] European Commission—European Commission. Available at: https://ec.europa.eu/commission/presscorner/detail/en/ip_20_17 (last accessed 1 September 2021).
113. International Labour Organisation, 2015. *Guidelines for a just transition towards environmentally sustainable economies and societies for all.* [online] Ilo.org. Available at: https://www.ilo.org/wcmsp5/groups/public/%2D%2D-ed_emp/%2D%2D-emp_ent/documents/publication/wcms_432859.pdf (last accessed 1 September 2021).
114. Canada, E., 2019. *Task Force: Just Transition for Canadian Coal Power Workers and Communities—Canada.ca.* [online] Canada.ca. Available at: https://www.canada.ca/en/environment-climate-change/services/climate-change/task-force-just-transition.html (last accessed 1 September 2021).

115. European Commission, 2021. *Just Transition Platform*. [online] ec.europa.eu. Available at: https://ec.europa.eu/info/strategy/priorities-2019-2024/european-green-deal/actions-being-taken-eu/just-transition-mechanism/just-transition-platform_en#timeline (last accessed 1 September 2021).
116. International Labour Organisation, 2015. *Guidelines for a just transition towards environmentally sustainable economies and societies for all*. [online] Ilo.org. Available at: https://www.ilo.org/wcmsp5/groups/public/%2D%2D-ed_emp/%2D%2D-emp_ent/documents/publication/wcms_432859.pdf (last accessed 1 September 2021).
117. Mulvey, K., 2020. *Just Transition Progress Report*. [ebook] Just Transition Commissioner. Available at: http://file:///C:/Users/O'Boyle%20family/Downloads/76825_2aedd89e-484c-4dea-a341-9d9036b46fd7.pdf (last accessed 1 September 2021).
118. Brennan, C., 2021. *Kieran Mulvey appointed Just Transition Commissioner for closure of peat-fired power stations*. [online] irishexaminer.com. Available at: https://www.irishexaminer.com/news/arid-30962706.html (last accessed 1 September 2021).
119. Climate Change Commission, 2021. *He Pou a Rangi—Climate Change Commission*. [online] Climatecommission.govt.nz. Available at: https://www.climatecommission.govt.nz/who-we-are/our-story/ (last accessed 1 September 2021).
120. Just Transition Commission, 2021. *Just Transition Commission: A National Mission for a fairer, greener Scotland—gov.scot*. [online] Gov.scot. Available at: https://www.gov.scot/publications/transition-commission-national-mission-fairer-greener-scotland/pages/6/ (last accessed 1 September 2021).
121. ACTU, 2021. *The Need for a Just Transition*. [online] Actu.org.au. Available at: https://www.actu.org.au/our-work/climate-change/the-need-for-a-just-transition [Accessed 22 April 2021].
122. Government of Canada, 2018. *Information archivée dans le Web | Information Archived on the Web*. [online] Publications.gc.ca. Available at: http://publications.gc.ca/collections/collection_2019/eccc/En4-361-2019-eng.pdf (last accessed 1 September 2021).
123. Government of Canada, 2018. *Information archivée dans le Web | Information Archived on the Web*. [online] Publications.gc.ca. Available at: http://publications.gc.ca/collections/collection_2019/eccc/En4-361-2019-eng.pdf (last accessed 1 September 2021).
124. Commission on Growth, Structural Change and Employment, 2019. *Final Report*. [online] Bmwi.de. Available at: https://www.bmwi.de/Redaktion/EN/Publikationen/commission-on-growth-structural-

change-and-employment.pdf?__blob=publicationFile&v=3 (last accessed 1 September 2021).

125. Commission on Growth, Structural Change and Employment, 2019. *Final Report.* [online] Bmwi.de. Available at: https://www.bmwi.de/Redaktion/EN/Publikationen/commission-on-growth-structural-change-and-employment.pdf?__blob=publicationFile&v=3 (last accessed 1 September 2021).
126. The Presidency Republic of South Africa, 2020. *Presidential Climate Change Coordinating Commission appointed.* [online] http://www.thepresidency.gov.za/press-statements/presidential-climate-change-coordinating-commission-appointed. Available at: http://www.thepresidency.gov.za/press-statements/presidential-climate-change-coordinating-commission-appointed (last accessed 1 September 2021).
127. ARC, 2021. *ARC's History and Work in Appalachia—Appalachian Regional Commission.* [online] Appalachian Regional Commission. Available at: https://www.arc.gov/arcs-history-and-work-in-appalachia/ (last accessed 1 September 2021).
128. European Commission, 2020. *Just Transition Platform.* [online] ec.europa.eu. Available at: https://ec.europa.eu/info/strategy/priorities-2019-2024/european-green-deal/actions-being-taken-eu/just-transition-mechanism/just-transition-platform_en#about-the-just-transition-platform (last accessed 1 September 2021).
129. The European Commission, 2021. *Press corner.* [online] European Commission—European Commission. Available at: https://ec.europa.eu/commission/presscorner/detail/en/ip_20_17 (last accessed 1 September 2021).

CHAPTER 5

Just Transitions Around the World

Abstract This chapter examines from a regional perspective the just transition to a low-carbon economy. Having assessed what the just transition involved in previous chapters, here the progress of countries is examined with a particular focus on government initiatives across the world in the following regions: Asia-Pacific, Middle East, Africa, Europe, Latin America, the Caribbean and North America. In particular, their specific plans for a just transition are assessed and how countries within each region are planning to develop a just transition policy.

Keywords Just transition • Asia-Pacific • Middle East • Africa • Europe • Latin America • The Caribbean • North America

5.1 Introduction[1]

This chapter examines the global progress of the just transition to a low-carbon economy (JTLCE). The term "just transition" has different meanings to different sectors and stakeholders within society; however, we are moving towards a harmonisation of its meaning. While, historically the term was used to secure social justice for workers, it now has evolved and has a broader goal for achieving justice within society and specifically as society moves to a low-carbon society.[2]

R. J. Heffron, *Achieving a Just Transition to a Low-Carbon Economy*, https://doi.org/10.1007/978-3-030-89460-3_5

The JTLCE aims to decarbonise the economies of the world and reduce harmful carbon dioxide emissions as a means of limiting global warming to 2° C or less by 2050. By reducing global warming, the global population hopes to limit further adverse effects of climate change which are only predicted to worsen as the planet heats up. Currently the worst affected regions are often home to the world's poorest and most vulnerable populations. It is inherently unjust that these regions now are affected disproportionately by carbon dioxide emissions as a result of more prosperous developed regions.

Therefore the JTLCE is a global issue in many aspects. It seeks to protect the health of the entire planet for future and current generations; it seeks to bring social, environmental and economic justice to the many not just the few; and most importantly it will take a global effort to achieve a universally just transition. Hence, in this chapter, the progress of the JTLCE is analysed across the different regions around the world and captures the movement at a critical time for the scholarship community.

5.2 Where Is the "Just Transition" Happening: Region by Region?

5.2.1 *Asia-Pacific*

Australia is in the top 20 global emitters of CO_2 and is the second largest exporter of coal in the world yet lacks national policies regarding the just transition. Despite signing the Paris Agreement, the government continues to back coal.[3] Even with frequent and severe climate-related natural disasters, the state's economic dependence on coal makes a transition complex. Opposition parties propose forming a Just Transition Authority yet also remain committed to coal.[4] Every Australian state has zero emission targets and in some cases renewable targets;[5] however, worker's unions are the main advocates demanding a just transition.[6,7] This includes calls for the creation of decent jobs in green energy and the reskilling of workers.[8]

India is also a heavily coal-dependent ESCAP state struggling with the just transition.[9] However ambitious renewable energy (RE) targets set in India's NDC indicate a will for the JTLCE.[10] The government's approach to the JTLCE is fragmented and incoherent, meaning the transition in India is mainly driven by NGOs and international programmes such as the recent “2030 Roadmap for India-UK future relations”. Despite being on track to miss the RE targets set in the NDC, India’s current energy pathway is still compatible with keeping global warming below 2° C.[11] Indonesia too depends heavily on coal and has previously been criticised for an unambitious NDC.[12] However, a recent announcement from the state-owned electricity distributor Perusahaan Listrik Negara (PLN) pledging to stop building coal plants by 2023 and only add renewable energy after this date suggests a JTLCE may be underway.[13]

China,[14] Georgia,[15] Azerbaijan[16] and the Philippines[17] are notably progressive states in the region. In the last decade, these countries have committed to successful respective development programmes that have seen significant economic, climate and environmental development.[18] Bhutan is another notable state in the region as the only carbon-negative country in the world.[19] The “Blue Pacific Big Ocean States”[20] are further examples of extremely vulnerable ESCAP states that lack a transition policy. In these nations, the International Labour Organisation (ILO) and local trade unions are the leading advocates for a JTLCE.[21]

Overall Regional Issues

The Asia-Pacific region is comprised of many states at various stages of economic, social and environmental development.[22] Several international initiatives aid JTLCE development such as the “Climate Investment Funds’ (CIF) Scaling Up Renewable Energy Program in Low-Income Countries” and investment programmes from the World Bank, Asia Development Bank (ADB) and European Bank for Reconstruction and Development (EBRD).

Demand for a “just transition” in the Asia-Pacific region appears to be uneven and largely insufficient. Trade unions such as the ILO and International Trade Union Confederation (ITUC) are critical in campaigning for a just transition and advocating on behalf of workers’ rights. Demands for better environmental policies are increasing in size and scope across the region in discussions on a just transition. There are vast differences in progress towards a just transition across the ESCAP region.

5.2.2 *The Middle East*

The Gulf Cooperation Council (GCC) is a key player in furthering the just transition in the Middle East.[23,24,25] Many countries in the Middle East Region are working towards national development strategies to promote

SD by 2050.[26] Despite this, in real terms, development has largely focussed on economic and social welfare often at the expense of environmental welfare and the energy transition.

The UAE[27] is the most advanced country in the region regarding the transition to a low-carbon economy. Although currently most energy is supplied by fossil fuels in the region, research indicates the kingdom is on track to reach its ambitious targets[28] set out under the UAE Future roadmaps that plan for up to 2021, 2021–2030 and 2030–2117.[29,30] The UAE has been successful through creating a stable policy environment,[31] continuous investment in renewable projects and successful use of policy instruments such as low power purchase agreement prices secured for projects.[32,33] Although this is positive in terms of the energy transition, the question remains over whether this transition is "just" as the kingdom has been accused of several human rights violations.[34]

Growing populations and subsequent pressure on the energy supply have greatly encouraged the development of RE in Jordan,[35,36] Oman,[37,38] and Egypt.[39,40] Jordan has made true progress towards achieving a *just* transition.[41] Iraq is on track to become the world's fourth largest oil producer by 2030, and as this state is central to the future world oil supply, it has a complicated relationship with a JTLCE.[42] Until recently Iraq's position on the JTLCE had been obscured by war and conflict in the region, resulting in a lack of data. However, in 2019 steps were made towards achieving SD and the just transition.[43,44] Saudi Arabia has an ambitious national development strategy (NDS) in "Saudi Vision 2030", yet recent analysis suggests the state is falling short on its commitments.[45]

Overall Regional Issues

Common themes of the NDS's of the region include aims to ensure each respective state's continued development to rank among the top developed nations in the world, to diverge and divest oil-based economies and to establish and/or increase the share of renewables in each country's energy mix (in many cases to gain energy independence from neighbouring states). Several states focus on increasing energy efficiencies and metering before introducing RE. Due to a high concentration of OPEC countries and hydrocarbon-dependent states, there is a clear pay-off between development needs and protecting the environment in the most developing countries of the region.

5.2.3 *Africa*

Algeria could become a key player in the global energy transition due to its size, natural resources and location.[46] However, despite ambitious 2030 transition energy targets, policy tools and financial mechanisms currently in use present barriers to an energy transition.[47,48] Djibouti's "Vision 2035" development strategy aims to increase the total share of renewable energy in the energy mix to 87–100% by 2035.[49] Unfortunately, in 2017 only 14.5% of TFEC[50] was met by renewable energy[51] due to an insufficient policy framework.[52]

Ethiopia has ambitious energy policy with a balanced focus on environment, climate and economy which suggests a truly just transition may be made in this state. Under the "Climate-Resilient Green Economy (CRGE)[53]", Ethiopia aims to become a lower-middle-income country by 2025 without increasing GHG emissions. This strategy falls under the Government's Ten-Year Perspective Plan (2020–2030)[54] which is supported by the Homegrown Economic Reform Agenda.[55]

Renewable energy accounted for approximately 80%[56] of Kenya's total energy supply in 2018.[57] Under the "Vision 2030"[58,59] NDS Kenya aims to meet 100% of energy demand with renewables by 2030. This includes plans to further diversify the energy mix, including more solar, wind and geothermal resources.[60] Nigeria is the largest economy in Africa[61] and continues to follow the National Renewable Energy and Energy Efficiency Policy (NREEEP)[62] and the SE4ALL Action Agenda to increase electricity access and the share of renewable energy in the energy mix by 2030.[63] Rwanda and Senegal are amongst the other African nations using SE4ALL-AAs to set ambitious energy targets also for 2030.

South Africa (SA) was the only country[64] to mention the just transition in its initial 2015 NDC.[65] Additionally, the 2020 Presidential Climate Change Coordination Commission's (PCCCC)[66] mandate to coordinate South Africa's just transition also encourages a JTLCE in SA. SA plans to decrease the state's heavy dependence on coal through increased shares of renewables[67] by 2030 yet has recently contradicted itself by signing long-term fossil fuel purchase agreements.[68]

Tunisia has ambitious plans to reach 30% renewable penetration by 2030; dynamic policy-making such as allowing businesses to produce their own energy has installed 6% of national energy production capacity via renewables.[69] Lastly in Zimbabwe, recent publication of the National

Renewable Energy Policy[70] has seen a national target of 26.5% renewable energy in the electricity mix[71] by 2030.

Overall Regional Issues

Low levels of electrification characterise the energy transition in Africa. States with limited electricity access are looking to renewable energy as an accessible and cost-effective means of facilitating development. Whether this is a "just" transition will depend on how policies include the welfare of the populations in the future and how the benefits of development are spread amongst the population. Although it is common for development strategies to acknowledge the role increased energy access and low-carbon energy will play in furthering socio-economic development, the term "just" transition is uncommon.

5.2.4 *Europe*

Under the European Green Deal (EGD),[72] the European Union (EU) aims to achieve a truly just energy transition with mechanisms such as The European Green Deal Investment Plan,[73] Just Transition Mechanism[74,75] and Just Transition Fund,[76] aiming to become the first carbon-neutral continent by 2050. The EGD furthers the ambitions of the Energy Union Strategy (EUS)[77] to facilitate a transition; the "Clean Energy for All Europeans" package (2019)[78] is the EU's most recent tool to implement the EUS.[79] The package enables a JTLCE by introducing a new energy market designed to absorb high penetrations of renewable energy, mandating 10-year integrated National Energy and Climate Plans (NECPs)[80] for 2021–2030 from each member state[81] and establishing an energy efficiency directive.

In Southeast Europe, the Energy Community Treaty[82] has been facilitating the energy transition in the contracting parties[83] since 2018. Signed in 2005, the treaty established the Energy Community to create a pan-European energy market that operates on the same legal framework for countries in the EU, Southeast Europe, the Black Sea region and beyond. The Energy Community Treaty has grown to incorporate climate and sustainability legislation,[84] and in 2018, the contracting parties signed the Wachau Manifesto into agreement which stated "the beginning of the end of the coal and lignite era" (Energy Community Secretariat, 2018, p1).[85]

Thanks to these policies, most European countries have the framework to achieve a JTLCE; however, some transitions are more genuine than others. Sweden leads the world in creating a low-carbon society.[86,87] Norway's policies[88] and progress seem comparatively advanced as shown in Fig. 2, yet Norway's achievements are undermined by the fact it remains one of the largest oil exporters in the world.[89] Norway uses the proceeds of petroleum wealth to "cancel out" emissions from oil by funding environmental projects overseas.[90] In this way, many EU countries depend on developing countries to "cancel out" emissions or for carbon-intense industries. This must be considered in future policy-making if global emissions are to be reduced.

Overall Regional Issues

Europe is uniquely successful[91] in its approach to achieving a JTLCE as much of the region is united under the EU. EU directives and frameworks create a uniform policy-making approach to the transition across member states which create favourable conditions for investment and private-public relationships in the region. Other factors contributing towards Europe's progress towards a JTLCE include nearly 100 years of development with no restrictions on fossil fuels, general political stability, a low percentage of economies dependent on exploiting fossil fuel resources and a comparatively mature and just political infrastructure.

5.2.5 Latin America

Several countries in the Latin American region are making promising steps towards JTLCE. Uruguay over the last two decades has transformed from importing large quantities of fossil fuels and electricity to meeting 98%–100% of electricity demand from RE. Building strong private-public sector relations, creating a stable regulatory environment and using policy tools such as fixed term feed-in tariffs can largely account for the success of the government's transition.[92]

Paraguay is another commendable country in the region. Currently the state meets 100% of electricity demand with hydropower resources and plans to meet 60% of energy demand with renewables by 2030, following the National Development Plan 2014–2030.[93] In 2018 Paraguay had the highest per capita surplus in renewable electricity production and the 5th highest production of RE per capita in the world.[94] Brazil is operating the National Energy Plan (PNE) 2050 designed by the Ministry of Mines and Energy of Brazil (MME) which aims for a renewable energy penetration of

45% by 2030,[95] aiming to maintain 2019 levels of low-carbon energy sources which accounted for 46% of Brazil's primary energy.[96,97]

Chile was the first country in South America to introduce a carbon tax and has created a strong infrastructure to support renewable projects through quotas, auctions and metering systems.[98] Chile has developed the "National Energy Policy 2050", which includes a broad set of energy efficiency goals for 2035 and 2050 and aims that RE will meet at least 60% of electricity demand by 2035 and 70% by 2050. Recent analysis suggests that the state's policy tools align with these targets and the updated 2020 NDC.[99] The NDC recognises the importance of the social, economic and environmental development and focuses on ensuring development is just in all senses. The NDC commits to developing a "Strategy for Just Transition" by 2021, which has yet to be released.[100] A just transition is complex in Chile due to a natural abundance of rare earth resources that are essential to develop RE globally, yet mining counts for 37% of electricity consumption in Chile.[101] Lithium mining is a focal point in the JTLCE for Chile and Bolivia as both countries have ambitious transition policies and yet must continue to exploit rare earth elements for the development of RE technologies.

Bolivia[102] is working under the Bolivia Electric Plan 2020–2025 which aims to increase energy access to 100% of the population and increase the share of renewable energy in the national energy mix.[103] This is guided by an ambitious NDC[104] which states Bolivia will increase the share of renewables in the electricity grid to 81% by 2030, compared to 39% in 2010.

Argentina has facilitated the beginning on a just energy transition through the introduction of the RenovAr programme. The programme aims to produce 20% of Argentina's electricity from renewable energy source by 2025 by creating a new market in Argentina for private investment in renewable energy.[105,106] Going forward Argentina must turn the energy transition into a just transition by including workers, promising jobs and creating the socio-economic development to support the transitioning population.[107,108]

5.2.6 *The Caribbean*

Antigua and Barbuda has committed to becoming net zero by 2050 and is currently working on a second NDC to confirm this target.[109] The government has acknowledged the need for a just transition for workers and intends to set just transition targets in the 2021–2025 NDC.[110] The

Bahamas has significant investment in renewable energy. Although the national energy policy does not mention a JTLCE by name, it does include a policy focus on "equity" and the fair distribution of benefits from the JTLCE.[111] Social and economic equity is also a focus in the Barbados Growth and Development Strategy which indicates the intention to host a truly just transition to green energy.[112]

Grenada has submitted an ambitious second NDC which aims to cut emissions by 40% by 2030. While the NDC does not mention a "just" transition, it outlines that Granada will take a holistic approach to low-carbon development that follows the principles of JTLCE.[113] In 2020, Jamaica also submitted a second, more ambitious NDC that doubles the island's climate ambition. Although the NDC highlights Jamaica's commitment to sustainable development, the document does not consider a just or equitable transition as a means of achieving the island's goals.[114] Cuba's updated NDC also submitted in 2020 covers the period 2020–2030 and shows increased ambition in reducing emissions, increasing shares of renewable energy and more intense forestry, land use and agriculture targets. The NDC does not mention sustainable development, a just transition or any principles of the JTLCE.[115]

Trinidad and Tobago (TT) is the largest oil- and gas-producing nation in the Caribbean. TT's NDC focusses on reducing emissions by 30% the combined power generation, industry and transport sectors by 2030.[116] Whilst the NDC does not mention the JTLCE, TT is currently drafting a national energy policy which embodies the principles of JTLCE including initiatives to promote education and awareness of RE and the reskilling of workers affected by the transition.[117]

Fifteen Caribbean states[118] are members of CARICOM[119] which has established the Caribbean Community Climate Change Centre (CCCCC).[120] The CCCCC is responsible for several initiatives that facilitate a just transition and build resilience to climate change across the region. This includes programmes such as the "2019–2023 Intra-ACP GCCA+ Programme in The Caribbean: Enhancing Climate Resilience in CARIFORUM Countries"[121] and the "Small Island Developing States (SIDS) Sustainable Energy Initiative".[122]

5.2.7 *North America*

Under President Biden, climate policy in America is undergoing radical changes.[123] Since being sworn in on January 20, 2021, Biden has re-entered the USA into the Paris Climate Agreement, published detailed executive orders[124] and targets[125] on how the USA will fight climate change and reduce emissions and established instruments such as the Climate Change Support Office (CCSO).[126] In April, the USA released an ambitious NDC;[127] however a complex political position limits the policy tools available to the administration.[128] The publication of the American Jobs Plan[129,130] has provoked unanimous descent from the Republican party thus illustrating the difficulties policy-making in the USA currently faces.[131]

Canada has made strong policy[132,133,134] and financial commitments[135] to becoming carbon-neutral by 2050. This includes the formation of a Just Transition Task Force.[136] There is a clear conflict between the exponentially increasing climate action in Canada and the deluge of hydrocarbon extraction and export taking place there. As the fourth largest oil producer in the world,[137] Canada must be held accountable for the emissions that the fossil fuels they export will eventually produce. For global emissions to fall, we must not put those who profit off hydrocarbon wealth to further their own social, economic and environmental agendas on a pedestal.

Energy and climate policy in Mexico are sadly fast transforming to favour fossil fuels over low-carbon energy solutions.[138] In December 2020 Mexico submitted a new NDC[139] which was found to lower Mexico's climate ambition and transparency.[140] Mexico has also been accused of engineering policy proposals to re-establish the state-owned, fossil fuel-dependent, energy and electricity companies' market dominance over renewable energy. Current policies encourage regulators and distributors to favour state-owned fossil fuel companies over renewable energy supplies, severely damaging what was once an emerging global leader in renewable energy markets.[141]

Politics threatens to prevent the JTLCE in the U.S. and in Mexico, whilst Canada could be accused of greenwashing their energy and climate policies. An economic dependence on fossil fuels prevents North America from a true rapid transition towards a low-carbon future, particularly in Mexico and Canada. The U.S.A. must find a quick method to achieve bipartisan agreement on the U.S.'s stance on climate change if policies are to succeed in the future. Canada's success relies on the international community demanding accountability for the volumes of oil produced by the state. The international community must also hold Mexico accountable for its recent back sliding and prioritisation of fossil fuels.[142]

5.3 Timelines for a Just Transition by World Region

5.3.1 Asia-Pacific

The Asia-Pacific region is currently not on track to achieve a just transition before 2030 or perhaps even 2050. Currently economic development depends heavily on carbon-heavy industries that historically have not supported climate or environmental justice. Research suggests the ESCAP region is not on track to meet any of the 17 SD Goals (SDG). Without intervention, the region may achieve less than 10% of the goals,[143] 14 of which are linked to achieving a just transition.[144] Figure 5.1 shows the regions progressing towards the 2020 milestone as of 2020 for each SDG including regressions on goals 13 and 14.

Australia is far away from meeting its 2030 NDC targets which are themselves weak and unambitious. Data has shown that Australia's greenhouse gas pollution will be 41 million tonnes higher in 2030 than in 2005.[145] Contrastingly India is on track to meet its 2030 targets and fulfil its NDC. Policy commitments such as the "2030 Roadmap for India-UK future relations" shows that India is committed to having a transition well underway by 2030 despite commitment issues with coal. Indonesia is currently not on track to meet its weak NDC;[146] however recent commitments from the state-owned electricity distributor PLN to stop building coal-fired power plants suggests there may still be hope for the JTLCE in Indonesia.[147]

China and the Philippines are some of the few states within the region that have clear and realistic timelines to transition to a low-carbon economy. China has an ambitious yet realistic plan to become carbon-neutral by 2060, whilst the Philippines aims to reduce emissions by 70% before 2030. Both countries appear to be largely on track due to necessary and drastic transformative policies on fossil fuel consumption; most notably the Philippines became the first Southeast Asian region to put a moratorium on new coal.

Additionally, Turkey,[148] Russia[149] and Vietnam[150] stand out as ESCAP states that are unlikely to meet any form of just transition without significant intervention and replanning. It should be noted that this seems to be a counterintuitive move on behalf of these states as the region is already feeling significant effects of climate change; since 1970 temperatures in Russia have been rising at 2.5 times the global average, and Vietnam is widely regarded to be one of the most vulnerable countries in the world to rising sea levels and monsoon rains.

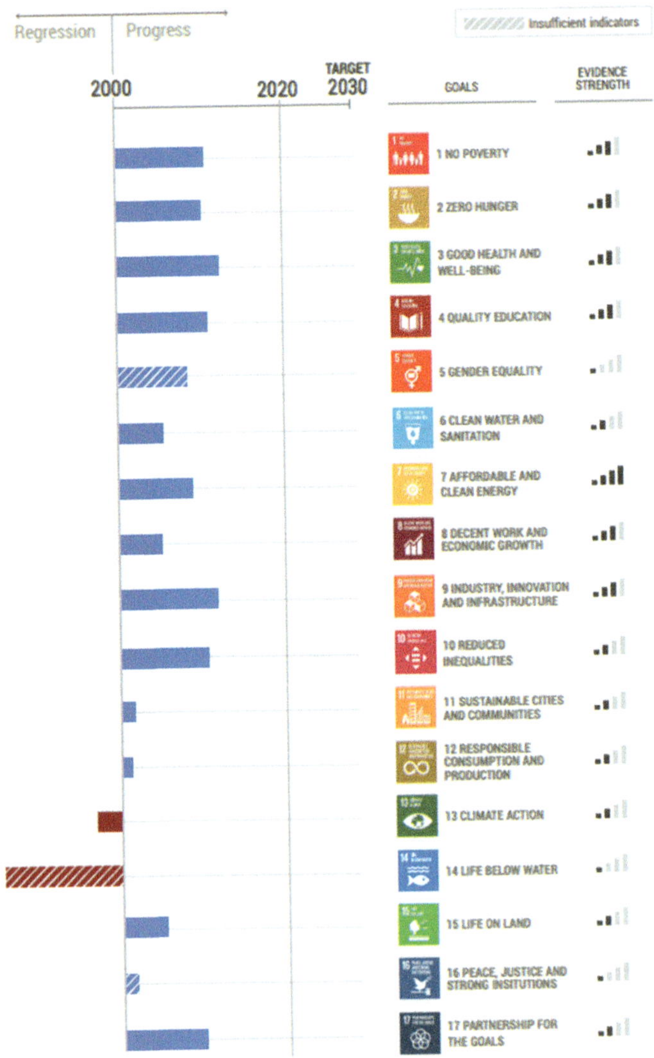

Fig. 5.1 Progress towards SDG 2020 milestone. (Source (UNESCAP, 2021))

5.3.2 *Middle East*

Despite members of the GCC and other countries in the region having NDS's that highlight 2030 as the year for achieving many significant elements of the JTLCE, the UAE is the only country currently on track to meet such targets. Although in the last decade the region has thrived in terms of social and economic development, the high concentration of hydrocarbon-dependent states creates a complicated relationship with environmental (climate) development.

Countries that have renewable deployment have already fallen behind their targets as is the case with Egypt,[151] Israel[152] and Kuwait.[153] In some cases, targets can be viewed as having negligible environmental impact when compared with the state's hydrocarbon activity as is the case of Bahrain and Qatar. In their NDS's, Israel[154] and Qatar[155] refer to gas as a clean fuel which is highly debatable and will significantly slow progress towards achieving a JTLCE. Even the UAE which is regarded as the most successful transition in the region has plans to incorporate "clean coal" into the energy mix which is another highly debatable term.[156] Continued reliance on fossil fuels will slow down the progress of the whole region.

In Saudi Arabia dependence on oil prevents a rapid JTLCE; there are influential actors who feel that an energy transition should be a multigenerational event to protect the economy.[157] The same can be said for other OPEC countries in the region including Kuwait, Iraq and Iran. Progress in Iraq has continued steadily in the last 12 months through a series of Memorandums of Understanding (MoU) between the state and various international organisations that will focus on social,[158] economic[159] and environmental[160] development. Establishing a timeline for JTLCE however remains difficult as although significant aid has been secured through MoUs RE is only mentioned once in "Iraq vision for sustainable development 2030".[161]

Conflict in some areas of the region such as Lebanon, Palestine, Syria and Yemen prevents a just transition from happening in these locations, although this has not prevented some governments from setting NDS's with energy targets.[162] NGOs[163] are active in these regions providing RE as a means of low-cost, accessible energy to rural communities, hospitals and areas where the national grid is unreliable. However, for a truly just transition to take place, significant political, social and economic reforms are necessary.

5.3.3 *Africa*

The just transition in Africa is tied inextricably to achieving socio-economic development. Timelines set in the national goals of these developing countries are especially vulnerable to delays as the country's infrastructure is often not resilient enough to absorb setbacks and maintain progress on the goals.[164] In Africa, there are several international actors investing in developing countries to create the infrastructure and facilities necessary to abate this situation. Notable donors include the African Development Bank,[165] Power Africa,[166] The People's Republic of China[167] and Sustainable Energy for All[168] (SE4ALL). SE4ALL has worked with 20 African countries[169] to develop individual SE4ALL Action Agendas. Each agenda provides a national policy framework for the SD of the energy sector in each state by 2030.[170]

Financially unsustainable energy subsidies and policy incoherence continue to prevent Algeria from achieving a low-carbon economy[171] before 2030.[172] In Djibouti, technical difficulties[173] with operation and deployment of renewables combined with weak private-public relations stall progress towards the 2035[174] targets.[175] Ethiopia continues to grow economically,[176] successfully staying on track to meet 2025[177] and 2030[178] targets allowing a transition towards a middle-income low-carbon economy.[179]

Kenya and Tunisia are steadily working towards their 2030 targets[180,181] as their share of renewable energy and energy policy[182,183] continues to grow.[184,185] In Nigeria construction of renewable projects is underway; however progress lags[186] the national targets.[187] The early stages of development projects in Zimbabwe look promising. Lastly in Zimbabwe, since the publication of the National Renewable Energy Policy, the government has approved 1GW of solar projects with loans from the world bank.[188] In 2019, renewable energy accounted for only 2.2% of the final energy mix in South Africa.[189] Combined with SA's recent 20-year commitment to fossil fuel purchase agreements, it does not seem likely that a just transition will happen in SA before 2030 or soon after.

Achieving a just transition implies a level of development to transition away from fossil fuels. In many African countries, there will only be sustainable development; the "just" elements of such development will be at the discretion of each individual nation. A small minority of countries in Africa have realistic plans for a just transition before 2030, and even fewer are on track to meet these goals. However, with investments and international aid, there is the potential for low-income countries in the region to skip the transition all together and achieve rapid sustainable development.

5.3.4 *Europe*

Europe is the only region endeavouring to achieve a unified transition before a specified date. The EU has historically led the world in climate policy with innovations such as the Emissions Trading Scheme (EU ETS)[190] and the Renewable Energy Directive (RED)[191] that have been essential to reducing emissions and fighting climate change in the last two decades. Now under the European Green Deal,[192] the EU aims to have undergone the transition to a low-carbon economy becoming carbon-neutral by 2050.[193] The Just Transition Mechanism has been designed to ensure the transition is fair and equitable.[194] The combined effects of the Clean Energy Package[195] on EU states and the Wachau Manifesto[196] on the contracting parties of the Energy Community would see widespread decarbonisation of the economy in most European states by 2030 if successful.

Great disparities remain between the levels of decarbonisation individual European states are aiming at. Germany,[197] Switzerland,[198] France[199] and the UK[200] have ambitious energy targets that aim to achieve carbon neutrality by 2050; however, targets are insufficient in eastern European countries that have a dependency on coal such as Bulgaria[201] and Poland.[202] The Scandinavian countries are widely perceived to lead the JTLCE in Europe,[203] yet as Denmark and Norway continue to export oil, the integrity of the JTLCE in these nations must be called into question. Despite Norway's[204,205] prodigious investments in environmental projects overseas, there is no direct correlation between the benefits of these projects and offsetting the negative effects borne from the burning of oil they export. Reforesting the rainforests of South America today will not protect the populations of Bangladesh, Indonesia or Vietnam from climate-induced floods in coming years.[206] Sweden[207] continues to lead the JTLCE in Europe with an ambitious target of net zero by 2045[208] as it achieves decarbonisation of space heating and electricity sectors whilst maintaining population and economic growth.

Despite differences in ambition across EU nations, a recent report on the cumulative impact of all 27 NECPs found that the EU is set to surpass its 2030 climate targets.[209] This has given cause for the European Commission to re-examine legislation to consider extending the targets.

This achievement is worthy of celebration; however, it is important to note that an independent analysis of current EU targets finds that they are not compatible with limiting global warming to 1.5° C.[210] It is concerning that the most advanced region in terms of achieving the JTLCE is still significantly far away from reaching goals agreed in the Paris Agreement.

5.3.5 *Latin America*

The energy transition in Latin America is uniquely complex. Geographically the region is extremely resource rich in hydrocarbon, rare earth and renewable resources. For the world to transition to a low-carbon economy, rare earth elements such as lithium, cobalt and copper must be mined, and yet this exact mining activity currently prevents a complete and/or rapid transition in many parts of the region. Additionally, there is resistance amongst workers in some countries who feel they are being left behind in the transition away from fossil fuels; this slows down progress towards a transition and highlights that in some regions the transition so far has not been just. This has led to the Trade Unions of the Americas (CSA-TUCA) developing an alternative just transition strategy for the region "Plataforma de Desarrollo de las Américas" (PLADA).[211] This strategy places the rights of the worker equal to the political, economic and environmental elements of the just transition.

Although almost every country in the region has a NDS, the transition will not be spread evenly. Ecuador,[212] Suriname[213] and Guyana[214] are still economically and/or energy dependent on fossil fuels and are unlikely to transition soon.[215,216] Peru and Venezuela are also still heavily dependent on oil and gas despite having high levels of hydropower in the energy mix; both state's energy transition policies have been criticised.[217] Despite this, several countries are setting and meeting ambitious energy targets.

Chile[218] and Colombia[219] maintain renewable energy targets for 2050 that are supported by strong policy infrastructures, making these targets achievable. Instruments such as energy auctions and imposed quotas on distributors like those implemented in Colombia from 2018 to 2021[220] have created stable conditions for investors in these countries. Brazil has made significant achievements in renewable energy; however, the current political situation threatens to outweigh previous progress. Brazil's decarbonisation could be pushed harder for a more rapid transition if policymakers lessened dependence on the fossil fuel industry.[221] Countries working towards the closer deadline of 2025 such as Argentina[222] and

Bolivia[223] have large amounts of work to do. Both countries have the natural resources and ambitious targets to achieve an energy transition; however, social factors such as worker's rights and the exploitation of natural resources must be considered if the transition is to be truly "just".[224]

5.3.6 *Caribbean*

The energy transition in many Caribbean states is restricted by financial barriers due to the developing country status of all members of the Caribbean Community (CARICOM).[225] Additionally, the COVID-19 pandemic has caused a drastic reduced tourist activities on many islands, several of whom counted on tourism as a main source of national income.[226]

Despite this an energy transition is underway in the region. Programmes established by the CCCCC make short- and medium-term provisions for the facilitation of a just transition and the sustainable development of a low-carbon economy in the region.[227] Programmes such as the "2009–2021 Regional Planning for Climate Compatible Development in the Region"[228] and capacity building programmes on each island[229] helps to create the infrastructure necessary for a just transition. Individual islands also have their own development strategies that often include energy transition and just transition policies.

Antigua and Barbuda want to put just transition mechanisms in the 2021–2025 NDC. The Bahamas National Energy Policy ensures equity and aims to facilitate a JTLCE in the period 2013–2033. Barbados Growth and Development Strategy also hopes to foster a JTCLE (perhaps just by another name) in the period 2013–2020. Grenada similarly does not mention a just transition by name, but by adopting JTLCE principles in its second ambitious NDC, Grenada hopes to have a just transition in the period 2030. Jamaica and Cuba have energy policies which encourage a low-carbon economy. These policies may be viewed as inherently just as they reduce emissions; however, little consideration has been given to mechanisms which would foster a just transition for communities, workers and industries affected by the transition.

Trinidad and Tobago also has an ambitious NDC regarding reducing emissions. The NDC similarly does not mention any other type of justice bar the implied effects reducing emissions will have; however Trinidad and Tobago is currently drafting a national energy policy that embodies the full scope of available just transition measures.

From this discussion, it can be surmised that there is largely the political willpower and policy mechanisms in place to facilitate a JTLCE in the Caribbean region by 2030. The physical realities of such a transition may be drawn out over a longer time span due to financial and in some cases political difficulties; however with increased international aid, the Caribbean could transition on or before target. Main challenges to overcome in the region include a heavy dependence on imported fossil fuels and low GDPs of the region further adversely affected by COVID-19.

5.3.7 *North America*

The JTLCE is different in each North American location; however it can be said that the region is not on track to achieve a just transition soon. The U.S. has historically failed to maintain commitments to energy and climate policies, and therefore many are cynical of Biden's plans. However, the breadth and pace of proceedings so far are promising indicators of how the U.S. may succeed if political agreement can be fostered across both sides of the political aisle. Canada has an admirable repertoire of policies and financial mechanisms to enable a JTLCE. Despite this Canada cannot truly transition and still maintain its status as the world's fourth largest oil producer. The international community must demand more accountability from Canada and other oil-producing states in this respect. Mexico must also be held to account by the international community for the back peddling on climate commitments in recent years; although faced with different development needs than Canada and the U.S. the current policy trajectory in Mexico is unacceptable.

There remains a possibility that the U.S. and Canada may achieve an energy transition on or before 2050. For the transition in the U.S. to be truly just, republicans must expand their definition of infrastructure to include societal elements such as education and care for the elderly thus allowing an energy transition and sustainable development where no one is left behind. In Canada, the greening of heavy polluting industrial sectors must not be the sole focus of future energy and emission reduction policies, especially before the appropriate technology is available. For Canada's transition to be truly just, there must be a transition away from fossil fuel extraction. Mexico is not on track to transition before 2050 and requires intervention if any transition is to happen at all.

A regional economic dependence on fossil fuels complicates the JTLCE in North America; however high degrees of policy ambition in Canada and the U.S. proves that there is a political will to make an energy transition happen that is fair and just. Mexico, Canada and the U.S. all have significant renewable energy resource potential, and this should aid the energy transition in each state. As renewable energy is an immediately available, cost-competitive source of low-carbon energy, North America must prioritise expanding renewable capabilities ahead of developing further fossil fuels, alternate fuels or CCUS technology resources that are still under development.

5.4 Final Reflections on the Just Transition with a Regional Perspective

The just transition must be flexible in relation to the development needs and economic structure of the country or region that is transitioning. What is just in one region may not be just in another, especially when comparing developed and developing countries or countries that are economically dependent on hydrocarbon production against those with divested economies. Although the fundamental principles of what makes a just transition "just" should remain the same in each region, the requirements necessary to achieve this justice will depend entirely on each location's unique circumstance. For example, developing countries in Latin America and Africa have still to make significant progress towards the JTLCE, yet it would be unjust to ask each region or indeed the individual countries to adopt identical transition measures immediately.

Further, policies implementing the JTLCE must respect the way in which national and local politicians and societies operate. The goal should be to help existing cultures transition instead of trying to create entirely new ones. Policy-makers' abilities to consider different social, political and economic characteristics will define the success (and failures) of just transition policy in the coming years.

The final point that can be drawn from this research is that the JTLCE is happening sporadically at different rates around the world, ranging from countries that have nearly completed the transition to countries that have yet to acknowledge a transition is needed. Additionally, no region in the world is on track to have a just transition on a timeline that will limit global warming to 2° C or lower before 2050.

The EU leads the world in creating and implementing policy that facilitates a JTLCE across an entire region, yet even here, not all member states have shown the ambition required to achieve EU targets. Important policy tools from the EU include the Just Transition Mechanism[230] and the European Green Deal.[231] Globally an important transition is being made towards RE; however progress is not happening quickly enough. Most regions of the world have targets to incorporate large-scale RE by 2030, increasing significantly in capacity by 2050.

However, as illustrated through the utopian scenario outlined in the IEA, recent publication of "Net Zero by 2050"[232] for the world to fulfil the Paris Agreement extreme universal change would have to be made in the coming days, weeks and months not over the coming decades. As previously established not all the transitions to RE are automatically just (or as just as they could be), and therefore the work still left to do is double fold. The first step to achieving a rapid JTLCE must be to rapidly increase the rate at which regions decarbonise economies and transition to RE. At the same time as this transition takes place, there must also be a rapid assessment of regional development policies around the world.

Appendix: Countries by Region

Asia-Pacific	Afghanistan; American Samoa; Armenia; Australia; Azerbaijan; Bangladesh; Bhutan; Brunei Darussalam; Cambodia; China; Cook Islands; Democratic People's Republic of Korea; Fiji; French Polynesia; Georgia; Guam; Hong Kong, China; India; Indonesia; Iran (Islamic Republic of); Japan; Kazakhstan; Kiribati; Kyrgyzstan; Lao People's Democratic Republic; Macao, China; Malaysia; Maldives; Marshall Islands; Micronesia (Federated States of); Mongolia; Myanmar; Nauru; Nepal; New Caledonia; New Zealand; Niue; Northern Mariana Islands; Pakistan; Palau; Papua New Guinea; Philippines; Republic of Korea; Russian Federation; Samoa; Singapore; Solomon Islands; Sri Lanka; Tajikistan; Thailand; Timor-Leste; Tonga; Turkey; Turkmenistan; Tuvalu; Uzbekistan; Vanuatu; Viet Nam
Middle East	Bahrain, Egypt, Iran, Iraq, Israel, Jordan, Kuwait, Lebanon, Oman, Palestine, Qatar, Saudi Arabia, Syria, The United Arab Emirates, Yemen
Africa	Algeria, Angola, Benin, Botswana, Burkina Faso, Burundi, Cameroon, Cabo Verde, Central African Republic, Chad, Comoros, Democratic Republic of Congo, Cote d'Ivoire, Djibouti, Equatorial Guinea, Egypt, Eritrea, Ethiopia, Gabon, Gambia, Ghana, Guinea, Guinea-Bissau, Kenya, the Kingdom of Lesotho, Liberia, Libya, Madagascar, Malawi, Mali, Mauritania, Mauritius, Morocco, Mozambique, Namibia, Niger, Nigeria, Rwanda, Sahrawi Arab Democratic Republic, Sao Tome and Principe, Senegal, Seychelles, Sierra Leone, Somalia, South Africa, South Sudan, Sudan, Kingdom of Swaziland, Tanzania, Togo, Tunisia, Uganda, Zambia and Zimbabwe

Europe	Albania, Andorra, Austria, Belarus, Belgium, Bosnia and Herzegovina, Bulgaria, Croatia, Cyprus, Czech Republic, Denmark, Estonia, Finland, France, Germany, Greece, Hungary, Iceland, Ireland, Italy, Latvia, Lithuania, Luxembourg, Malta, Monaco, Montenegro, Netherlands, North Macedonia, Norway, Poland, Portugal, Republic of Moldova, Romania, San Marino, Serbia, Slovakia, Slovenia, Spain, Sweden, Switzerland, Ukraine, United Kingdom of Great Britain and Northern Ireland
Latin America	Argentina, Bolivia, Brazil, Chile, Colombia, Ecuador, Guyana, Paraguay, Peru, Suriname, Uruguay, Venezuela
North America	Canada, Mexico, United States of America
Caribbean Islands	**Independent States**: Antigua and Barbuda, Bahamas, Barbados, Cuba, Dominica, Dominican Republic, Grenada, Haiti, Saint Kitts and Nevis, Jamaica, Saint Lucia, Saint Vincent and the Grenadines, Trinidad and Tobago

Notes

1. It is with express thanks to my Research Assistant Cara McKinney who worked on this chapter in 2021 with me and responded to all direction in assisting with the different drafts of the chapter.
2. Heffron, Raphael J. & McCauley, Darren 2018. What is the 'Just Transition'? *Geoforum*, 88, 74–77.
3. Through approving new coal mines and coal powered electricity generation.
4. Green, Fergus. 2019. Australia: caught between a 'just transition' and 'no transition'. Foundation of European Progressive Studies (FEPS).
5. Tracker, Climate Action. 2020l. *Australia* [Online]. Climate Action Tracker. Available: https://climateactiontracker.org/countries/australia/ (last accessed 1 September 2021).
6. (to protect the rights of workers who will be affected by the predicted eventual decline of coal).
7. Maher, Tony A New Approach in Australia to Just Transition. National President, Construction, Forestry, Mining and Energy Union, Australia.
8. Unions, Australian Council of Trade. *The Need for a Just Transition* [Online]. ACTU. Available: https://www.actu.org.au/our-work/climate-change/the-need-for-a-just-transition#:~:text=At%20a%20minimum%20Australia%20needs,the%20economies%20of%20impacted%20communities (last accessed 1 September 2021).
9. Although emissions per capita are below the global average, India is the third biggest CO_2 emitter in the world Singh, Rajesh Kumar. 2021. India to Double Down on Coal Projects Amid Climate Warnings. *Bloomberg*, 25.03.21.

10. IEA (2021), India Energy Outlook 2021, IEA, Paris https://www.iea.org/reports/india-energy-outlook-2021 (last accessed 1 September 2021).
11. Tracker, Climate Action 2020e. India. 22.09.20 ed.: Climate Action Tracker.
12. Indonesia, Republic of 2016. First Nationally Determined Contribution.
13. Lo, Joe. 2021. Indonesian utility pledges to stop building coal plants beyond existing pipeline. *Climate Home News* [Online]. Available from: https://www.climatechangenews.com/2021/05/11/indonesian-utility-pledges-stop-building-coal-plants-beyond-existing-pipeline/ (last accessed 1 September 2021).
14. Stern N and Xie C (2021) China's New Growth Story: Linking the 14th Five-Year Plan with its 2060 Carbon Neutrality Pledge. London: Grantham Research Institute on Climate Change and the Environment, London School of Economics and Political Science.
15. IEA (2020), Georgia 2020, IEA, Paris https://www.iea.org/reports/georgia-2020 (last accessed 1 September 2021).
16. Development, European Bank for Reconstruction and 2016. Life in Transition III. European Bank for Reconstruction and Development.
17. Institute, World Resource Philippines: A Whole-of-government Approach to Creating Green Jobs. World Resource Institute.
18. Azerbaijan, Ministry of Ecology and Natural Resources of the Republic of. *Ministry of Ecology and Natural Resources of the Republic of Azerbaijan* [Online]. Available: http://eco.gov.az/ (last accessed 1 September 2021).
19. Despite progressive environmental policies, there are still elements of the transition to be made in Bhutan. However, the state is due to graduate from the LDC category in 2023 and is improving socially and economically—Tracker, Climate Action. 2020j. *Bhutan* [Online]. Climate Action Tracker. Available: https://climateactiontracker.org/countries/bhutan/#:~:text=In%20its%20Nationally%20Determined%20Contribution,change%20and%20forestry%20(LULUCF). [Accessed 25.04.2021]. Youn, Soo 2017. Visit the World's Only Carbon-Negative Country. *National Geographic*. National Geographic, Affairs, United Nations Department of Economic and Social & Analysis, Economic 2018. Least Developed Country Category: Bhutan Profile. United Nations.
20. Cook Islands, Fiji, Kiribati, Marshall Islands, Palau, Papua New Guinea, Samoa, Solomon Islands, Tonga, Tuvalu and Vanuatu.
21. Organization, International Labour 2019. Green Jobs and a Just Transition for Climate Action in Asia and the Pacific. Thailand: ILO.
22. This paper will refer to the countries grouped under the United Nations Economic and Social Commission for Asia and the Pacific (ESCAP) as

belonging to the Asia-Pacific (see Appendix 1 for full list of countries and regions).

23. The council unites governments in Bahrain, Kuwait, Oman, Qatar, Saudi Arabia and the United Arab Emirates to improve security, economic well-being and social development in the region. In partnership with global organisations such as the World Bank, the council implements programmes that focus on improving human capital, environmental reforms, fiscal developments and policy reforms.
24. Bank, The World 2021. The World Bank in the Gulf Cooperation Council. *Where We Work*. 24.03.21 ed.: The World Bank.
25. The council was initially formed to strengthen interstate relations and development across the region; a full list of the council's objectives can be found here: https://www.gcc-sg.org (last accessed 1 September 2021).
26. Bahrain 'The Economic Vision 2030', 'Egypt's Vision 2030', Iran 'Vision 2025', 'Iraq Vision for Sustainable Development 2030', 'Israel 2050', 'Jordan 2025', Kuwait Vision 2035, Lebanon USAID Development Programme, Oman 'Vision 2040', Qatar 'National Vision 2030', 'Saudi Arabia Vision 2030'.
27. United Arab Emirates.
28. Scully, Jules 2021. UAE solar capacity to increase fourfold by end of 2025 thanks to 'robust' development pipeline. *PV Tech*. PV Tech.
29. Emirates, United Arab 2021. UAE Future. United Arab Emirates Government.
30. The plans are varied and complex including plans to establish the first habitable human settlement on Mars.
31. The UAE 'Energy Strategy 2050' has been fundamental; IRENA, United Arab Emirates Ministry of Energy & Industry 2017. UAE National Energy Strategy 2050. *Presentation for CEM Long Term Energy Scenarios*. IRENA.org: IRENA, United Arab Emirates Ministry of Energy & Industry.
32. Energy, Rystad 2021. It's raining solar panels in the UAE: Renewable capacity set to increase fourfold to 9 GW by end-2025. Rystad Energy.
33. Other achievements include establishing Masdar City, the first carbon-neutral, zero waste city in the region and the success of the Dubai Clean Energy Strategy 2050; (Masdar. 2021. *Masdar* [Online]. Available: https://masdar.ae/en/masdar-city/the-city [Accessed 03.05.21 2021], Administration, International Trade 2020a. United Arab Emirates Country Commercial Guide. International Trade Administration, Authority, Dubai Electricity & Water 2020. Renewable energy shaping the future of sustainability. Government of Dubai.)
34. (International, Amnesty 2020. UNITED ARAB EMIRATES 2020. 2020 ed.: Amnesty International.)

35. Jordan's share of electricity from renewables grew from almost zero in 2014 to around 20% in 2020. This is made possible through a framework of policies such as Jordan 2025: A National Vision and Strategy (2015) and subsequent Executive Development Programmes which turn the "Jordan 2025: A National Vision and Strategy" into shorter, more manageable action plans in three and four year cycles for constant and effective self-assessment and progress towards the end goal (Jordan, The Hashemite Kingdom of 2015. JORDAN 2025: A National Vision and Strategy. The Hashemite Kingdom of Jordan.
36. (IRENA 2021b. IRENA Report Identifies Policy Measures to Advance Jordan's Transition to Renewables. IRENA.), (Jordan, The Hashemite Kingdom of Jordan's Way to Sustainable Development: First National Voluntary review on the implementation of the 2030 Agenda. New York: High-Level Political Forum on Sustainable Development.)
37. In 2016 Oman ranked 2nd most sustainable country in the Arab Region. Under the NDS "Oman Vision 2040", Oman continues to use 5-year plans to guide progress. A key innovation of the Omani strategy is diversification of the economy to reduce the country's economic dependency on oil, aiming for more than 90% of non-oil GDP. The government has announced that it aims to have 30% of Oman's electricity demands produced through renewable means by 2030 and has developed several large- and small-scale projects in line with this goal.
38. Guirindola-Astolfi, Agnes 2017. SUSTAINABLE DEVELOPMENT IN OMAN: A TIMELINE. Sustainable Oman.(Administration, International Trade 2020c. Oman's Renewable Energy Projects. 04.27.20 ed.: International Trade Administration.) (Oman, Sultanate of 2013. OMAN VISION 2040. Oman: Sultanate of Oman.)
39. Egypt has the infrastructure and resources for a just energy transition; recent IRENA reports that RE could provide as much as 53% of Egypt's energy mix by 2030. The Egyptian government is working under an energy diversification strategy, known as the 2035 Integrated Sustainable Energy Strategy (ISES). ISES states Egypt's aim is producing 20% of its electricity using renewable sources by 2022 and 42% by 2035. There are also goals for wind to provide 14%, hydropower 2% and solar 25%.
40. Wärtsilä. 2020. Egypt's energy ambitions. Available: https://www.wartsila.com/insights/article/egypts-energy-ambitions [Accessed 04.05.21], IRENA 2018. Renewable Energy Outlook: Egypt. IRENA in collaboration with the Government of Egypt, as representated by NREA.
41. Jordan made significant progress towards achieving the Millennium Development Goals and has approached the SDGs with the same thoroughness and commitment; however it is clear that the unstable conditions in the region (namely the Syrian refugee crisis) pose serious threats

to development; Nations, United 2017. Jordan: Voluntary National Review. Sustainable Development Goals Knowledge Platform.

42. IEA (2019), *Iraq's Energy Sector: A Roadmap to a Brighter Future*, IEA, Paris https://www.iea.org/reports/iraqs-energy-sector-a-roadmap-to-a-brighter-future (last accessed 1 September 2021).
43. Including the creation of a Sustainable Development Goals (SDG) Committee and publication of the "Iraq vision for sustainable development 2030" development strategy.
44. Iraq Vision for Sustainable Development 2030. *In*: Planning, M. o. (ed.). Republic of Iraq, Iraq, United Nations. 2019. *Sustainable Development Goals (SDG) Committee inaugurated in Kirkuk* [Online]. United Nations. Available: http://www.uniraq.com/index.php?option=com_k2&view=item&id=10826:sustainable-development-goals-sdg-committee-inaugurated-in-kirkuk&Itemid=606&lang=en (last accessed 1 September 2021).
45. Daoud, Ziad. 2021. Scoring Saudi Arabia Vision 2030 Five Years After Launch. Available: https://www.bloomberg.com/news/articles/2021-04-28/scoring-saudi-arabia-vision-2030-five-years-after-launch-chart (last accessed 1 September 2021).
46. Algeria connects the "MENA region with both southern Europe and sub-Saharan Africa. Algeria is the wealthiest country in the region by GDP per capita, and the largest nation on the continent by land area. In addition, it enjoys more than 3000 hours of sunlight annually, giving it substantial solar potential, yet it ranks third in Africa in terms of emissions of carbon dioxide (C02), a powerful greenhouse gas" (Hochberg, 2020).
47. And therefore, a just transition cannot take place.
48. (Hochberg, Michael 2020. Algeria charts a path for renewable energy sector development. Middle East Institute.) (2020a. Renewable Energy and Energy efficiency Development Plan 2011–2030. November 2020 ed.: Climate Policy Database.)
49. Djibouti, Republic of 2014. Vision Djibouti 2035. Republic of Djibouti.
50. Total Final Energy Consumption.
51. IRENA Energy Profile Djibouti. Abu Dhabi: IRENA.
52. USAID. 2020. *Djibouti Power Africa Fact Sheet* [Online]. USAID. Available: https://www.usaid.gov/powerafrica/djibouti (last accessed 1 September 2021).
53. Ethiopia, The Government of the Federal Democratic Republic of 2011. Ethiopia's Climate-Resilient Green Economy—Green economy strategy. The Government of the Federal Democratic Republic of Ethiopia.
54. Ethiopia, The Government of the Federal Democratic Republic of 2021. Ethiopia 2030: The Pathway to Prosperity. *Ten Years Perspective Development Plan (2021–2030).*

55. Ethiopia, The Government of the Federal Democratic Republic of 2019. A Homegrown Economic Reform Agenda: A Pathway to Prosperity.
56. Including biomass.
57. Kenya, Republic of 2018. National Energy Policy.
58. The "National Energy Policy" (2018) is instrumental in supporting this vision.
59. Kenya, Republic of 2007. Kenya Vision 2030.
60. Kenya, Republic of 2018. National Energy Policy.
61. IEA. 2018a. *Nigeria Key Energy Statistics* [Online]. IEA. Available: https://www.iea.org/countries/nigeria (last accessed 1 September 2021).
62. Nigeria, Federal Republic of 2015. National Renewable Energy and Energy Efficiency Policy (NREEEP). *In*: Power, M. o. (ed.).
63. Federal Republic of Nigeria, Sustainable Energy for All 2016. Sustainable Energy for All Action Agenda (SE4ALL-AA). *In*: Efficiency, I.-M. C. o. R. E. a. E. (ed.). SE4ALL.
64. Institute, World Resource. *South Africa: Strong Foundations for a Just Transition* [Online]. Available: https://www.wri.org/just-transitions/south-africa (last accessed 1 September 2021).
65. Africa, Government of South 2015. South Africa's Intended Nationally Determined Contribution (INDC).
66. Africa, Republic of South 2020. Cabinet approves climate change, emissions reduction and waste management plans.
67. Africa, Republic of South 2019. Integrated Resource Plan. *In*: Department, E. (ed.).
68. Cotterill, Joseph. 2021. South Africa under fire over energy plan. *Financial Times*, 10.05.21.
69. Administration, International Trade 2020b. Tunisia Country Commercial Guide. 13.07.20 ed.
70. Zimbabwe, Republic of 2019. National Renewable Energy Policy. *In*: Development, M. o. E. a. P. (ed.).
71. Excluding hydro which currently provides around 57%—Independence, Off Grid Energy. 29.11.19 2019. Zimbabwe to Benefit from Pumped Hydroelectric Energy Storage. Available from: https://www.offgridenergyindependence.com/articles/18975/zimbabwe-to-benefit-from-pumped-hydroelectric-energy-storage (last accessed 1 September 2021).
72. Commission, European 2019a. The European Green Deal. Brussels.
73. Commission, European 2020c. Financing the green transition: The European Green Deal Investment Plan and Just Transition Mechanism. Brussels.
74. Commission, European. *The Just Transition Mechanism: making sure no one is left behind* [Online]. Available: https://ec.europa.eu/info/strat-

egy/priorities-2019-2024/european-green-deal/actions-being-taken-eu/just-transition-mechanism_en (last accessed 1 September 2021).

75. The EGD has several notable mechanisms including The European Green Deal Investment Plan and Just Transition Mechanism which direct investment and infrastructure into areas such as private-public partnerships and transitioning worker's roadmaps to ensure that the transition is full and just.
76. Commission, European 2020a. Commission welcomes the political agreement on the Just Transition Fund.
77. Commission, European 2015. Communication From The Commission To The European Parliament, The Council, The European Economic And Social Committee, The Committee Of The Regions And The European Investment Bank: A Framework Strategy for a Resilient Energy Union with a Forward-Looking Climate Change Policy. European Commission.
78. Commission, The European 2019b. Clean energy for all Europeans. Luxembourg: The European Commission.
79. The package focuses on five key areas: energy security, a fully integrated internal energy market, energy efficiency, climate action and decarbonising the economy and fostering research, innovation and competitiveness inside the energy market.
80. (following a common structure). Note that the transparent nature of the NECPs encourages a high standard of commitment to be maintained regarding progress towards each nation's targets.
81. (outlining how each nation will achieve their individual targets on all aspects of the EUS)
82. Community, The Energy 2015. Treaty establishing Energy Community. Athens, Greece.
83. Albania, Bosnia and Herzegovina, Georgia, Kosovo, Moldova, Montenegro, North Macedonia Serbia and Ukraine
84. Team, The Energiewende. 2018. Energy Community sets stage for clean energy transition in South-East Europe. *Energy Transition The Global Energiewende* [Online]. Available from: https://energytransition.org/2018/07/energy-community-sets-stage-for-clean-energy-transition-in-south-east-europe/#more-17577 (last accessed 1 September 2021).
85. Secretariat, Energy Community 2018. The Wachau Manifesto. The Energy Community.
86. Sweden, The Government of 2020. Sweden's Integrated National Energy and Climate Plan. *In*: Infrastructure, T. M. o. (ed.).
87. Under the "Integrated National Energy and Climate Plan" (2016), Sweden aims for a net zero economy by 2045. Sweden has achieved eco-

nomic growth while imposing high carbon taxes and halved energy generated from fossil fuels by implementing a market-based approach to the energy market.

88. Norway, Government of 2019. Norway's National Plan related to the Decision of the EEA Joint Committee *In*: Environment, N. M. o. C. a. (ed.).
89. (with no plans to decrease dependence on fossil fuels)—Petroleum, Norwegian. 2021. *Production Forecasts* [Online]. Norwegian Petroleum. Available: https://www.norskpetroleum.no/en/production-and-exports/production-forecasts/ (last accessed 1 September 2021).
90. Pinker, Dr. Annabel 2020. Just Transitions: a comparative perspective. The James Hutton Institute & SEFARI Gateway.
91. Renewable energy penetration is high, the energy market is becoming profitable for low-carbon energy, and there are mechanisms to ensure the creation of green jobs across the region.
92. Bertram, Rebecca. 2020. *Uruguay, Latin America's Renewable Champion* [Online]. Available: https://energytransition.org/2020/01/uruguay-latin-americas-renewable-champion/ (last accessed 1 September 2021), Administration, International Trade 2020d. Uruguay Country Commercial Guide. 14.10.20 ed.: International Trade Administration.
93. IEA. 2018 *Paraguay Key Energy Statistics* [Online]. IEA. Available: https://www.iea.org/countries/paraguay [Accessed], Edge, Leading 2018 Paraguay: pioneer in renewable and hydroelectric energy supply. *Leading Edge Guides.* Leading Edge.
94. Development, United Nations Conference on Trade and 2018. UNCTAD Multi-year Expert Meeting on TRADE, SERVICES AND DEVELOPMENT Water and Sanitation, Energy and Food-related Logistics Services, Country paper: PARAGUAY. UNCTAD.
95. Energy, Ministry of Mines and 2020. Plano Nacional de Energia 2050. Brazil: Ministério de Minas e Energia. Empresa de Pesquisa Energética.
96. Roser, Hannah Ritchie and Max 2020b. Brazil: Energy Country Profile. 2020 ed.: Our World in Data.
97. Moving forward further renewable development in Brazil will move away from hydro (28.7% of energy in 2019) and focus on solar energy as 90% of unrealised yet "exploitable" hydropower potential in the country highlighted in the previous PNE2030 was in the Amazon region. (Environment, Grantham Research Institute on Climate Change and the. *National Energy Plan 2030 (PNE 2030)* [Online]. National Energy Plan 2030 (PNE 2030). Available: https://climate-laws.org/geographies/brazil/policies/national-energy-plan-2030-pne-2030 (last accessed 1 September 2021).

98. Bersalli, Germàn 2019. CHILE, ENERGY: An emerging key actor in the renewable energy arena. Chile: Climate Chance.
99. Tracker, Climate Action 2020d. Chile. 30.07.20 ed.: Climate Action Tracker, ibid. ibid.
100. Chile, Government of 2020a. Chile's Nationally Determined Contribution Update 2020. *In*: Sustainability, C. o. M. f. (ed.).
101. Bersalli, Germàn 2019. CHILE, ENERGY: An emerging key actor in the renewable energy arena. Chile: Climate Chance.
102. Several additional programmes have been launched to support these ambitions such as the Programa Electicidad para Vivir con Dignidad (PEVD, Electricity Program for Living with Dignity) and The Access and Renewable Energy Project that work with international actors such as the World Bank, the International Development Association, Germany's KfW and the InterAmerican Development Bank. Bank, The World. 2020. *Increasing Access to Electricity and Renewable Energy in Bolivia* [Online]. The World Bank. Available: https://www.worldbank.org/en/results/2020/10/15/increasing-access-to-electricity-and-renewable-energy-in-bolivia (last accessed 1 September 2021).
103. IEA. 2018d. *Bolivia Key Energy Statistics* [Online]. IEA. Available: https://www.iea.org/countries/bolivia (last accessed 1 September 2021).
104. BOLIVIA, THE PLURINATIONAL STATE OF 2015. INTENDED NATIONALLY DETERMINED CONTRIBUTION FROM THE PLURINATIONAL STATE OF BOLIVIA. THE PLURINATIONAL STATE OF BOLIVIA.
105. This was done through a series of extremely successful renewable energy auctions facilitated by the World Bank and IFC who also helped create an infrastructure between the government and private sector.
106. Bauza, Vanessa 2017. A New Dawn: Argentina Taps Into Its Renewable Energy Potential. International Finance Corporation.
107. Currently workers feel they are being left behind in the transition, and without the safety of green jobs, they resist the transition. (Castro, Nazaret 2020. Argentina at an energy crossroads: the role of trade unions in building a just transition. Equal Times.)
108. Recent reports find that with the falling cost of renewable technologies and the progress made by the state, Argentina has the capabilities to surpass their agreed upon target and could update it to renewable energy penetration 28%–30% in 2025 and 38%–43% in 2030. (Leonardo Nascimento, Marie-Jeanne Kurdziel, Hanna Fekete, Markus Hagemann, & Vivero, Gustavo de 2020. Decreasing costs of renewables—Implications for Argentina's climate targets. New Climate Institute.)

109. Charlene Watson, MIchai Robertson, Ameera Ramdin, Courtnae Bailey 2020. Assessment and Overview of Climate Finance Flows: Antigua And Barbuda 2014–2017. UNFCCC.
110. Black-Layne, Diann 2020. NDC Targets for the Just Transition of the Workforce. *Climate Dialogues 2020—Enhancing Capacity and Understanding on Assessment of Impacts on Implementation of Response Measures to Facilitate Economic Diversification and Just Transition.* Department of Environment.
111. Housing, Ministry of the Environment and 2013. The Bahamas National Energy Policy 2013–2033. Ministry of the Environment and Housing.
112. Affairs, Ministry of Finance and Economic 2013. Barbados Medium-Term Growth and Development Strategy 2013–2020. *In*: The Economic Affairs Division, M. o. F. a. E. A. (ed.).
113. Grenada, Government of 2020. Grenada Second Nationally Determined Contribution.
114. Jamaica, Government of 2020. Update of Nationally Determined Contribution of Jamaica.
115. Cuba, Republic of 2020. Summary of the first NDC Updated 2020–2030.
116. Tobago, Government of Trinidad and 2015. Intended Nationally Determined Contribution (iNDC) Under the United Nations Framework Convention on Climate Change.
117. Industries, Ministry of Energy and Energy. *Brief on Renewable Energy and Energy Efficiency* [Online]. Government of Trinidad and Tobago. Available: https://www.energy.gov.tt/our-business/alternative-energy/renewable-energy/ (last accessed 1 September 2021).
118. Antigua and Barbuda, Bahamas, Barbados, Belize, Dominica, Grenada, Guyana, Haiti, Jamaica, Montserrat, Saint Lucia, St Kitts and Nevis, St Vincent and the Grenadines, Suriname, and Trinidad and Tobago.
119. Community, Government of the Caribbean 2001. Revised Treaty of Chaguaramas Establishing the Caribbean Community Including the CARICOM Single Market and Economy.
120. CARICOM 2002. Agreement Establishing the Caribbean Community Climate Change Centre (CCCCC). Belize City, Belize: CARICOM.
121. CCCCC. *2019–2023 Intra-ACP GCCA+ Programme In The Caribbean: Enhancing Climate Resilience In CARIFORUM Countries* [Online]. Available: https://www.caribbeanclimate.bz/blog/2021/04/07/2019-2023-intra-acp-gcca-programme-in-the-caribbean-enhancing-climate-resilience-in-cariforum-countries/ (last accessed 1 September 2021).
122. CCCCC. *2012–2016 SIDS DOCK* [Online]. Available: https://www.caribbeanclimate.bz/blog/2017/07/27/2012-2016-sids-dock/ (last accessed 1 September 2021).

123. From January 2017 to 2021, President Trump torpedoed America's climate commitments by leaving the Paris Climate Agreement, relaxing environmental regulations and increasing subsidies on fossil fuels.
124. Jr, President Joseph R. Biden 2021b. Executive Order on Tackling the Climate Crisis at Home and Abroad.
125. Room, The White House Briefing 2021a. Fact Sheet: President Biden Sets 2030 Greenhouse Gas Pollution Reduction Target Aimed at Creating Good-Paying Union Jobs and Securing U.S. Leadership on Clean Energy Technologies.
126. Jr, President Joseph R. Biden 2021a. Executive Order on the Establishment of the Climate Change Support Office.
127. Government, The United States of America 2021. The United States of America Nationally Determined Contribution.
128. Biden's administration is currently trying to achieve these goals by only implementing policies that do not have to get passed by congress as this has historically caused climate policy in the U.S. to fail: Heath, Ryan 2021. Biden's climate plan faces global skepticism. *Politico.*
129. Room, The White House Briefing 2021b. Fact Sheet: The American Jobs Plan.
130. Which is in many ways a roadmap to achieving a JTLCE in the U.S.
131. Gross, Samantha. 10.05.21 2021. Barriers to achieving US climate goals are more political than technical. *Brookings* [Online]. Available from: https://www.brookings.edu/blog/planetpolicy/2021/05/10/barriers-to-achieving-us-climate-goals-are-more-political-than-technical/ (last accessed 1 September 2021).
132. Canada has impressive infrastructure to ensure 2030 and 2050 targets are met: Canada, Government of 2020a. An Act respecting transparency and accountability in Canada's efforts to achieve net-zero greenhouse gas emissions by the year 2050. *In*: Canadian Parliament, H. o. C. (ed.). Ottawa: House of Commons of Canada.
133. Canada's budget and roadmap to climate neutrality: Canada, Government of 2020b. A Healthy Environment and a Healthy Economy. *In*: Canada, E. a. C. C. (ed.). Gatineau, QC.
134. In April Canada announced that it will join the U.S. in creating the "Net-Zero Producers Forum", a community for oil- and gas-producing countries to discuss how they can best achieve the JTLCE and net zero emissions by 2050: Canada, Government of 2021a. Canada Joins U.S. in Establishing Net-Zero Producers Forum. Ottawa: Natural Resources Canada.
135. Amongst other promises Canada has committed to investing $3 billion over 7 years through the Strategic Innovation Fund (SIF) into the Net Zero Accelerator Fund to implement rapid decarbonisation of the econ-

omy: Canada, Government of 2021b. Budget 2021: Building an Innovation Economy of the Future. Department of Finance Canada.

136. 2019a. *Task Force: Just Transition for Canadian Coal Power Workers and Communities* [Online]. Available: https://www.canada.ca/en/environment-climate-change/services/climate-change/task-force-just-transition.html (last accessed 1 September 2021).
137. N. Sönnichsen. 2020. *Oil industry in Canada—Statistics & facts* [Online]. Available: https://www.statista.com/topics/2963/canadian-oil-and-gas-industry/#topicHeader__wrapper (last accessed 1 September 2021).
138. Flannery, Nathaniel Parish 2021. Political Risk Analysis: Is Mexico Declaring War Against Clean Energy? *Forbes*.
139. Mexico, Government of 2020. Nationally Determined Contributions 2020 Update.
140. Tracker, Climate Action 2020b. CAT Climate Target Update Tracker Mexico. 30.12.20 ed.: Climate Action Tracker.
141. Herrera, Carolina. 23.04.21 2021. Mexico Falls Further Behind on Climate at Leaders Summit. *NRDC Expert Blog* [Online]. Available from: https://www.nrdc.org/experts/carolina-herrera/mexico-falls-further-behind-climate-leaders-summit-0 (last accessed 1 September 2021).
142. Haldevang, Amy Stillman and Max De. 2021. Energy Protectionism in Mexico Has Made Climate the Victim. *Bloomberg*, 09.01.21.
143. UNESCAP 2021. Asia and The Pacific SDG Progress Report. *In*: ESCAP, U. N. (ed.). SDG Gateway: United Nations ESCAP.
144. Development, European Bank for Reconstruction and. *What is a just transition?* [Online]. Available: https://www.ebrd.com/what-we-do/just-transition (last accessed 1 September 2021).
145. Stock, Petra. 2018. Let's Get Something Straight—Australia Is Not On Track To Meet Its Paris Climate Target. *Climate Council* [Online]. Available from: https://www.climatecouncil.org.au/australia-not-on-track-to-meet-climate-targets/?atb=DSA01b&gclid=Cj0KCQjwh_eFB-hDZARIsALHjIKc2SnAkbGNh5ryl8PDjmWdDbSAdbH4uDDMn-GUSrW414jrHQRNzpQuIaAqMxEALw_wcB (last accessed 1 September 2021).
146. Tracker, Climate Action 2020i. Indonesia. 22.09.20 ed.: Climate Action Tracker.
147. Lo, Joe. 2021. Indonesian utility pledges to stop building coal plants beyond existing pipeline. *Climate Home News* [Online]. Available from: https://www.climatechangenews.com/2021/05/11/indonesian-utility-pledges-stop-building-coal-plants-beyond-existing-pipeline/ (last accessed 1 September 2021).

148. Tracker, Climate Action. 2020f. *Turkey* [Online]. Climate Action Tracker. Available: https://climateactiontracker.org/countries/turkey/ (last accessed 1 September 2021).
149. Tracker, Climate Action. 2020h. *Russia* [Online]. Climate Action Tracker. Available: https://climateactiontracker.org/countries/russian-federation/ (last accessed 1 September 2021).
150. Tracker, Climate Action. 2020a. *Vietnam* [Online]. Climate Action Tracker. Available: https://climateactiontracker.org/countries/viet-nam/ (last accessed 1 September 2021).
151. Roser, Hannah Ritchie and Max 2020d. Egypt: Energy Country Profile. 2020 ed.: Our World in Data.
152. Despite commitments to increase the share of renewable energy in the energy mix to 10% by 2020, in 2019 renewable energy accounted for only 2.3% of the energy supply in Israel. (Roser, Hannah Ritchie and Max 2020c. Israel: Energy Country Profile. 2020 ed.: Our World in Data.)
153. Butler, Nick. 2019. Kuwait's action on green energy fails to match best intentions. *Financial Times*, 10.09.19.
154. Israel, State of 2018. Energy Economy Objectives for the Year 2030. *In*: Energy, M. o. (ed.). Jerusalem: State of Israel.
155. Qatar, State of 2008. Qatar National Vision 2030. *In*: Planning, G. S. F. D. (ed.). Doha.
156. Emirates, United Arab. 2019. *UAE Future 2030–2117* [Online]. Available: https://u.ae/en/more/uae-future/2030-2117 (last accessed 1 September 2021).
157. McQue, Katie 2020. Energy transition aside, Saudi Arabia to maintain oil exports 'for decades,' official says. *In*: Editor (ed.). S&P Global Platts.
158. Pineau, Miriam 2021. Building a peaceful Iraq: New MoU between UNDP Iraq and swisspeace signed Baghdad: Relief Web.
159. 2020c. DFC, Iraq Sign Memorandum of Understanding Supporting Investment in Development. Washington: U.S. International Development Finance Corporation.
160. 2020b. Iraq's environmental sustainability prioritized in new UNEP/UNDP agreement. Manama, Baghdad: UNDP.
161. Iraq Vision for Sustainable Development 2030. *In*: Planning, M. o. (ed.). Republic of Iraq.Iraq Vision for Sustainable Development 2030. *In*: Planning, M. o. (ed.). Republic of Iraq.Iraq Vision for Sustainable Development 2030. *In*: Planning, M. o. (ed.). Republic of Iraq.Iraq Vision for Sustainable Development 2030. *In*: Planning, M. o. (ed.). Republic of Iraq.Iraq Vision for Sustainable Development 2030. *In*: Planning, M. o. (ed.). Republic of Iraq.Iraq Vision for Sustainable Development 2030. *In*: Planning, M. o. (ed.). Republic of Iraq.Iraq Vision for Sustainable Development 2030. *In*: Planning, M. o. (ed.).

Republic of Iraq.Iraq Vision for Sustainable Development 2030. *In*: Planning, M. o. (ed.). Republic of Iraq.

162. ((RCREEE), Regional Centre for Renewable Energy and Energy Efficiency Yemen. RCREEE.)((RCREEE), Regional Centre for Renewable Energy and Energy Efficiency Syria. RCREEE.)(Conservation, Lebanese Center for Energy 2016. The National Renewable Energy Action Plan for the Republic of Lebanon 2016–2020. *In*: Water, M. o. E. a. (ed.). Lebanese Center for Energy Conservation.)
163. (Mistiaen, Veronique 2020. Yemen's 'microgrid girls' power community amid war and COVID-19. Aljazeera.)(Thayer, Jack 2021. Renewable Energy in Palestine. Borgen Project.)
164. Unfortunately, many setbacks encountered by these developing countries are caused or exacerbated by their lack of socio-economic development creating a cycle of poverty and underdevelopment.
165. Bank, African Development. *African Development Bank* [Online]. Available: https://www.afdb.org/en (last accessed 1 September 2021).
166. Africa, USAID Power. *Power Africa* [Online]. USAID. Available: https://www.usaid.gov/powerafrica (last accessed 1 September 2021).
167. Lema, Rasmus, Bhamidipati, Padmasai Lakshmi, Gregersen, Cecilia, Hansen, Ulrich Elmer & Kirchherr, Julian 2021. China's investments in renewable energy in Africa: Creating co-benefits or just cashing-in? *World Development*, 141, 105365.
168. All, Sustainable Energy for. *The SEforALL Africa Hub* [Online]. Available: https://www.se4all-africa.org/the-africa-hub/who-we-are/the-seforall-africa-hub/ (last accessed 1 September 2021).
169. Gambia, Cabo Verde, Angola, Benin, Burkina Faso, Cabo Verde, Cote d'Ivoire, Ghana, Kenya, Liberia, Nigeria, Mali, Mozambique, Rwanda, Sierra Leone, Swaziland, Tanzania, Togo, Uganda and Guine-Bissau.
170. Within this framework, several states have made impressive progress towards building a low-carbon energy sector including Rwanda, Senegal, Cape Verde and Uganda. However, infrastructure issues remain in other nations using the agendas (such as Burkina Faso, Gambia and Liberia) which prevents these states from meeting their goals. While the SE4ALL agendas do foster the development of a low-carbon economy in many nations, there are little to no mentions of "justice" or a "just" transition.
171. Haddoum, Saliha, Bennour, Hocine & Ahmed Zaïd, Toudert 2018. Algerian Energy Policy: Perspectives, Barriers, and Missed Opportunities. *Global Challenges*, 2, 1700134.
172. 2020a. Renewable Energy and Energy efficiency Development Plan 2011–2030. November 2020 ed.: Climate Policy Database.
173. Program, Energy Sector Management Assistance 2017. Just Below the Surface: 24/7 Renewable Energy for Djibouti. ESMAP.

174. Djibouti, Republic of 2014. Vision Djibouti 2035. Republic of Djibouti.
175. IRENA 2020. Scaling Up Renewable Energy Deployment in Africa. IRENA.
176. Ethiopia is one of only seven countries in the world predicted to maintain GDP growth of over 3% for the fiscal year 2020.
177. Ethiopia, The Government of the Federal Democratic Republic of 2011. Ethiopia's Climate-Resilient Green Economy—Green economy strategy. The Government of the Federal Democratic Republic of Ethiopia.
178. Ethiopia, The Government of the Federal Democratic Republic of 2021. Ethiopia 2030: The Pathway to Prosperity. *Ten Years Perspective Development Plan (2021–2030)*.
179. Tracker, Climate Action. 2020k. *Ethiopia* [Online]. Climate Action Tracker. Available: https://climateactiontracker.org/countries/ethiopia/current-policy-projections/ (last accessed 1 September 2021).
180. L'Énergie, Agence Nationale Pour La Maitrise De. *Our Vision* [Online]. ANME. Available: http://www.anme.tn/?q=fr/content/notre-vision (last accessed 1 September 2021).
181. Kenya, Republic of 2007. Kenya Vision 2030.
182. IEA. 2018b. *Tunisian Solar Plan (PST) 2010–2016* [Online]. Available: https://www.iea.org/policies/4936-tunisian-solar-plan-pst-2010-2016 (last accessed 1 September 2021).
183. Such as the Energy Act passed in 2019 to further regulate the energy market—Kenya, Republic of 2019. The Energy Act. Nairobi.
184. Spielkamp, Rüdiger. Solar Plan: to produce 30 per cent of electricity from renewable energy sources. *Giz* [Online]. Available from: https://www.giz.de/en/worldwide/60432.html 2021]. (last accessed 1 September 2021).
185. Janho, Roy. 18.11.20 2020. Renewable energy in Kenya: An examination of the legal instruments and institutional changes that successfully attracted foreign investment. *Clean Power Professionals Group* [Online]. Available from: https://energycentral.com/c/pip/renewable-energy-kenya-examination-legal-instruments-and-institutional-changes (last accessed 1 September 2021).
186. Dioha, Michael. 2018. Nigeria's Renewable Energy Policy: A Fantasy or Reality? *Renewable Energy World* [Online]. Available from: https://www.renewableenergyworld.com/solar/nigerias-renewable-energy-policy-a-fantasy-or-reality/#gref 2021]. (last accessed 1 September 2021).
187. (Federal Republic of Nigeria, Sustainable Energy for All 2016. Sustainable Energy for All Action Agenda (SE4ALL-AA). *In*: Efficiency, I.-M. C. o. R. E. a. E. (ed.). SE4ALL.) (Nigeria, Federal Republic of 2015. National Renewable Energy and Energy Efficiency Policy (NREEEP). *In*: Power, M. o. (ed.).)

188. Bellini, Emiliano 2020. Zimbabwe set for real solar growth. *PV Magazine*. PV Magazine.
189. Roser, Hannah Ritchie and Max 2020a. South Africa: Energy Country Profile. 2020 ed.: Our World in Data.
190. Parliament, The European 2003. Establishing a scheme for greenhouse gas emission allowance trading within the Community and amending Council Directive 96/61/EC.
191. Parliament, The European 2009. on the promotion of the use of energy from renewable sources and amending and subsequently repealing Directives 2001/77/EC and 2003/30/EC.
192. Commission, European 2019a. The European Green Deal. Brussels.
193. Union, European 2020. Long-term low greenhouse gas emission development strategies (LT-LEDS). Zagreb.
194. Commission, European. *The Just Transition Mechanism: making sure no one is left behind* [Online]. Available: https://ec.europa.eu/info/strategy/priorities-2019-2024/european-green-deal/actions-being-taken-eu/just-transition-mechanism_en (last accessed 1 September 2021).
195. Commission, The European 2019b. Clean energy for all Europeans. Luxembourg: The European Commission.
196. Secretariat, Energy Community 2018. The Wachau Manifesto. The Energy Community.
197. Germany, The Government of 2016. Climate Action Plan 2050. *In*: Federal Ministry for the Environment, N. C., Building and Nuclear Safety (BMUB) (ed.). Berlin.
198. Switzerland, The Federal Council of 2021. Switzerland's Long-Term Climate Strategy. *In*: Federal Department of the Environment, T., Energy and Communications (DETEC) (ed.).
199. France, The Government of 2020. Plan National Integre Energie-Climat De La France.
200. Kingdom, Government of the United 2020. The Ten Point Plan for a Green Industrial Revolution. *In*: Ministry of Business, E. I. S. (ed.).
201. Bulgaria sees coal as a baseload fuel and essential to electricity generation: Bulgaria, Republic of 2019. Integrated Energy and Climate Plan of the Republic of Bulgaria.
202. Poland aims to have 60% coal in the electricity mix in 2030: 2018. Energy Policy of Poland Until 2040. *In*: Energy, M. o. (ed.). Warsaw.
203. Forum, World Economic 2021. Fostering Effective Energy Transition 2021 Edition. World Economic Forum.
204. For balance Norway's state-owned energy company Equinor does have a roadmap to invest in low-carbon solutions and renewable energies and reach net zero by 2050; however Equinor and the Norwegian government have no plans to slow the production of oil or gas.

205. Equinor. *Our Climate Ambitions* [Online]. Equinor. Available: https://www.equinor.com/en/sustainability/climate.html (last accessed 1 September 2021).
206. Coca, Nithin. 2020. Flooded Asia: Climate change hits region the hardest. *Financial Times, Nikkei Asia*, 15 October 2020.
207. IEA 2019. Sweden 2019 Review. *Energy Policies of IEA Countries.* IEA.
208. Sweden, The Government of 2020. Sweden's Integrated National Energy and Climate Plan. *In*: Infrastructure, T. M. o. (ed.).
209. Commission, European 2020b. Driving forward the green transition and promoting economic recovery through integrated energy and climate planning Brussels.
210. Tracker, Climate Action 2020g. EU. 22.09.20 ed.
211. Americas, Trade Union Confederation of The 2019. PLADA X 2030 Agenda. TUCA.
212. (IEA. 2018e. *Ecuador Key Energy Statistics* [Online]. IEA. Available: https://www.iea.org/countries/ecuador [Accessed 05.05.21 2021].) (IRENA 2015b. Renewable Energy Policy Brief Ecuador. Abu Dhabi: IRENA.)
213. IEA. 2018c. *Suriname Key Energy Statistics* [Online]. IEA. Available: https://www.iea.org/countries/suriname [Accessed 05.05.21 2021].
214. Initiative, Energy Transitions 2020. Guyana Energy Snapshot. U.S. Department of Energy.
215. Ecuador is currently working under the National Plan for energy efficiency 2016–2035 which plans energy efficiency over the period 2016–2035. Approximately 31% of energy came from renewables in 2019 with the rest supplied by oil (66.42%) and gas (3.06%). Although this is a relatively high percentage, the National Plan for Good Living 2013–2022 (PNBV-SENPLADES 2013–2017) previously set the target of reaching 60% renewable energy generation capacity by 2017, and therefore it could be seen that Ecuador is a long way of the energy transition.
216. IRENA 2015a. Renewable Energy Policy Brief Venezuela. Abu Dhabi: IRENA.
217. 2019b. Peru targets investment in renewable energy. *The Report, Peru 2019* [Online]. Available: https://oxfordbusinessgroup.com/analysis/looking-sun-work-under-way-attract-capital-renewable-energy#:~:text=Peru's%20energy%20development%20strategy%20intends,of%20all%20installed%20generation%20capacity. (last accessed 1 September 2021). (Bellini, Emiliano 2019. Venezuela plans its first utility scale PV projects. *PV Magazine*.)
218. Chile, Gobierno de 2020b. Energia 2050 Política Energética de Chile. *In*: Energía, M. d. (ed.).) (IEA (2018), Energy Policies Beyond IEA

Countries: Chile 2018 Review, IEA, Paris https://www.iea.org/reports/energy-policies-beyond-iea-countries-chile-2018-review (last accessed 1 September 2021).
219. 2019c. PLAN ENERGÉTICO NACIONAL 2020–2050. *In*: (UPME), L. U. d. P. M. E. (ed.). Bogotá: La Unidad de Planeación Minero Energética (UPME).
220. IRENA 2021a. Renewable energy auctions in Colombia: Context, design and results. Abu Dhabi: IRENA.
221. Tracker, Climate Action. 2020c. *Brazil* [Online]. Climate Action Tracker. Available: https://climateactiontracker.org/countries/brazil/ (last accessed 1 September 2021).
222. Corporation, International Finance RenovAr (Argentina): Scaling 'Express Edition'. IFC.
223. BOLIVIA, THE PLURINATIONAL STATE OF 2015. INTENDED NATIONALLY DETERMINED CONTRIBUTION FROM THE PLURINATIONAL STATE OF BOLIVIA. THE PLURINATIONAL STATE OF BOLIVIA.
224. Castro, Nazaret 2020. Argentina at an energy crossroads: the role of trade unions in building a just transition. Equal Times.
225. Community, Government of the Caribbean 2001. Revised Treaty of Chaguaramas Establishing the Caribbean Community Including the CARICOM Single Market and Economy.
226. This limits finances available for the transition.
227. CCCCC. *CCCCC Mission Statement*: [Online]. Available: https://www.caribbeanclimate.bz/about-us/mission/ (last accessed 1 September 2021).
228. CCCCC 2009. Climate Change and the Caribbean: a regional Framework For achieving development resilient to Climate Change (2009–2015). Belmopan, Belize: Caribbean Community Climate Change Centre (CCCCC).
229. CCCCC. *Green Climate Fund* [Online]. Available: https://www.caribbeanclimate.bz/green-climate-fund-map/ (last accessed 1 September 2021).
230. Commission, European. *The Just Transition Mechanism: making sure no one is left behind* [Online]. Available: https://ec.europa.eu/info/strategy/priorities-2019-2024/european-green-deal/actions-being-taken-eu/just-transition-mechanism_en (last accessed 1 September 2021).
231. Commission, European 2019a. The European Green Deal. Brussels.
232. IEA 2021. Net Zero by 2050. Paris: IEA.

CHAPTER 6

The Elements of the Just Transition Within International Institutions

Abstract This chapter analyses the impact and influence of the just transition within and from international institutions. It explores their role utilising the key forms of justice from the Just Framework and highlights how these institutions are working on: distributive justice, procedural justice, restorative justice, recognition justice and cosmopolitanism justice. A range of NGOs and development banks are those prime international institutions that are already exploring the just transition issues, and it seems certain that they are set to continue their work in this area. It is development banks who have the potential to play a more significant role given their ability to finance the just transition, and it is of importance they are now working in this area.

Keywords Just transition • International institutions • Just Framework • Restorative justice • Cosmopolitan justice

It is with express thanks to my two Research Assistants Jamie Hunt and Marie Tritz who worked on this chapter in 2021 with me and responded to all direction in assisting with the different drafts of the eventual chapter.

R. J. Heffron, *Achieving a Just Transition to a Low-Carbon Economy*, https://doi.org/10.1007/978-3-030-89460-3_6

6.1 Introduction

The evolution of the role of international institutions has prompted a new role in international governance.[1] International institutions have become an important feature on the international scene by providing support to their members for their own development. The scope of international institutions differs from one to another, but most international institutions are specialised which allows for more targeted and efficient actions including reports, projects and funding. Even though a large majority of international institutions are specialised, they can also work on similar projects as most policies today are interlinked to climate actions, employment and economic growth.

In this chapter, the focus is on international institutions and on how they focus on the key justice elements—that is, distributive, procedural, restorative, recognition and cosmopolitan. Here we assess from these perspectives how these international institutions are contributing to policy-making, research activity and wider societal impact on the just transition to a low-carbon economy. For a list of the institutions and brief description, see Appendix.

6.2 Procedural Justice, International Institutions and the Just Transition

Overall, the above international institutions have demonstrated procedural justice in their efforts towards the just transition through a framework perspective as well as financial and reform-based progress. Moreover, emphasis on the continued compliance with the Paris Agreement is another instance where procedural justice has been executed.

6.2.1 Creating Access to Finance

Some institutions especially development banks and economic institutions have tried to foster the just transition by developing new conditions for loans to convince their member countries to make the shift in their policies. A good example is Remap options which redirect investments of around US $5 billion annually from fossil fuel technologies into renewables.[2] The growing place of just transition has also called upon further investments which are now conditioned to the specific criteria.[3]

Additionally, the European Investment Bank (EIB) has also adopted this approach. In one of its latest staff papers, the EIB outlines clear eligibility criteria for funds.[4] The grant resources can take different forms like interest rate subsidies requiring concessional financing sources, for example, under IMF debt sustainability requirements or for sovereign lending targeting sectors with high economic and social returns which will certainly contribute a just transition.[5] Another form is "investment grants: coupled with loan funds to target asset funding in projects that aim to deliver high economic and social impact, including for climate action".[6]

Acknowledging the impacts of climate change on investments, banks have been frontrunners in the implementation of a just transition by interlinking loans with climate imperatives. One of these incentives has been the investments in digital technologies for climate change adaptation and resilience like InsurTech.[7] This initiative touches upon procedural justice but also distributive justice as out of the $80 million Blue Orchard InsuResilience Investment Equity Sub-Fund, the EIB committed $20 million to equity participation.[8] This initiative invests in insurers that invest in and use technology and in digital technology companies developing solutions for the insurance industry, and a recent example is Inclusive Guarantee, a leading microinsurance broker in West Africa that promote socially inclusive insurance products and now focusing on West African farmers and index-based crop insurance.[9] Looking back at the latest achievements of microinsurance, it has developed through three phases. The first phase focuses on community-based organisations like cooperatives, trade unions and faith groups. Second phase integrates new technologies to increase the distribution like mobile network.[10] Lastly, "the microinsurance landscape has been transitioning away from mobile sales as insurers explore other customer aggregation channels, such as digital platforms and apps".[11]

6.2.2 *Consumer Rights*

Another illustration of procedural justice could be found regarding consumers rights. In some part of the world, consumers rights have been infringed as they have not been able to compete with the biggest companies. Taking the example of the energy transition, the less concurrent companies are present on the market, the bigger companies will increase the prices worsening energy access. Therefore, procedural justice encourages government to focus on consumer perspective and their rights to justice but also on the affordability of energy services.[12] In the past, Indonesian consumers have been left behind by the government as the coal industry had more resources than mere citizens to influence governments' decisions.[13]

6.3 Distributive Justice, International Institutions and the Just Transition

Distributive justice concerns "justice in the distribution of benefits and burdens to individuals, or consists in the balancing of the competing claims persons make on the benefits that are up for distribution".[14] Distributive justice aims to include everyone in the equation, and it directly echoes with the idea of a just transition. It can be found in most of the reports drafted by international institutions. IRENA highlighted in its 2016 report for ASEAN that "many developing regions would benefit from pursuing green growth. ASEAN Member States are no exception. The region is highly vulnerable to climate change, posing a particular threat to farmers and coastal communities".[15]

6.3.1 Focusing on Distributive Issues Across Vulnerable Groups

By referring to the vulnerability of the region, IRENA also reaffirms the need for inclusive measures to ensure that no one is left beyond. Moreover, it should be noted that in most reports, distributive justice is intertwined with recognition. Thus, most examples related to distributive justice might also be linked to recognition since it acknowledges the vulnerability of a specific group. For example, Cambodia and Myanmar plans to enable full access to electricity of all their populations, respectively, by 2020 and 2030.[16] Electricity accessibility is a redundant theme especially in Africa and Asia as it also concentrates a big share of world's poor people, so a just transition should be for everyone but also at a fair price. Consequently, electricity prices should not deepen equalities between citizens especially vulnerable and poor people that need to be considered in the reform process.[17] Reflecting on the energy transition, distributive justice should guarantee a fair distribution of energy and secure the development of democratic ownership of energy resources triggers fairness.[18]

Everyone needs to be considered in the just transition: migrant workers and refugees but also other more vulnerable groups that might not be visible.[19] The EIB has, for instance, highlighted the existence of a significant risk for greater social inequality because of the pandemic for more vulnerable groups.[20] Most of reports dealing with just transition have now integrated the pandemic consequences. As of today, the pandemic has revealed the fragility and the inadaptability of countries system but also accelerated inequalities. In fact, economic challenges and social distancing measures are likely to "trigger social unrest and destabilisation in sub-Saharan African countries".[21] "The African Development Bank anticipates a GDP contraction of 1.7% to 3.4% in 2020, and estimates that nearly 50 million Africans could be pushed into extreme poverty by the economic consequences of the pandemic. Poorly diversified economies are expected to be most seriously affected".[22]

One good example of distributive justice is the plan of Costa Rica. Their National Decarbonization Plan applies to ten sectors but also includes labour strategies for a just transition.[23] Building the just transition across all sectors but also integrating all the different actors will ensure that no one is left behind. Hence, distributing all benefits from the just transition illustrates only one part of the distributive side of just transition; it should also favour vulnerable groups like workers.

6.3.2 *Initiating Social Dialogue with Workers Around Distributive Justice*

Besides the inclusion of vulnerable people, the just transition will also require the development of new job skills. The Inter-American Development Bank and International Labour Organization have highlighted that "of the 22.5 million jobs created in Latin America and the Caribbean by 2030 under the decarbonization scenario, 13.5 million jobs fall into the medium-skill category, while 8.2 million jobs will be created for low-skill workers and 820,000 jobs for high-skill workers".[24] Yet, low-skill occupations will experience the highest proportion of job gain and loss.[25] As low-skill jobs will increase, workers would need access to a fair income to avoid more inequalities. Improving working conditions and income will distribute more equally the benefits of the just transition.

The multiplication of social dialogue would ensure that no one is left behind. As argued in one report, the concerns and needs of workers but also communities will help build trust and forge consensus for the transition.[26]

6.4 Restorative Justice, International Institutions and the Just Transition

Restorative justice looks after any injustice caused that "should be rectified and it focuses on the need for enforcement of particular laws (i.e., energy sites should be returned to former use, hence waste management policy and decommissioning should be properly done)".[27] It can either ensure that those left behind will be protected, that law compliance will be strengthened or that any inequality will be addressed.

6.4.1 *Restoring Education Access*

As already argued, the pandemic has indeed put more people at risks and disturbs their everyday life more than we can possibly measure. Yet, it is also a chance to learn from it and accelerate the just transition. In the case of Africa, reports have focused on human capital development. The EIB highlighted the educational gap between African countries especially regarding digital skills. "For instance, the World Bank reports that only 50% of African countries include computer skills in their curriculum, compared to 85% elsewhere".[28] To fill the gap, African countries will need to count on public-led initiatives (e.g. Rwanda's smart classroom) but also on private initiatives led by companies, local players and even foreign companies with the *IBM Digital—Nation Africa* to teach future learners an IBM cloud-based learning platform to host free educational programmes for five years.[29]

Another good example can be found in Argentina and Guyana. On the one hand, the Guyana Energy Agency had anticipated the future needs in the energy sector in the context of a low-carbon economy and drafted a ten-year plan "Human Resource Development Plan" which will not only update electrical engineering but also develop certification for standardising skills for renewable energies.[30] On the other hand, Argentinian government has focused on green jobs. To develop green jobs in the country, provincial governments like the one from Santa Fe promoted these jobs by developing trainings, certification for green job competencies, occupational standards as well as improving working conditions.[31] Following the success of the programme called "Green Jobs—Caring for the Planet Provides Work" (*Empleos Verdes—Cuidar el planeta da trabajo* in Spanish), it is likely that further programmes will be developed in the following years to restore a balance between workers.[32]

6.4.2 *Reforming Financial Support*

Another form of restorative justice is the development of financial support. Yet, it is easier said than done. As pointed out by ILO, the just transition raises the question of "how to protect vulnerable populations when reforming energy subsidies so as to reduce emissions without hurting the poor".[33] Price incentives can represent a significant barrier to adopting carbon-free technologies in the energy and transport sectors so governments should consider how best to spend the money. The reform of financial schemes does not qualify the whole system; it, however, needs to consider the additional impacts a project might have on a community. For example, Menengai Geothermal Power Station project aimed at providing power to about 500,000 households and 300,000 businesses and is supported by the French Development Agency, the EIB and the Government of Kenya in 2011.[34] Since then, the project has even attracted new investors like the United States Agency for International Development (USAID) and Japan International Cooperation Agency (JICA) with a "long-term goal of developing 465MW of geothermal steam equivalent. A total of 47 geothermal wells, with a potential of 169MW, have been drilled so far by 2019".[35]

Another good example is the Just Transition Mechanism (JTM) developed by the EU.[36] The aim of the JTM is to support "the regions and sectors that are most affected by the transition because they depend on fossil fuels or carbon-intensive processes. [...] It will also strive to protect the citizens and workers most vulnerable to the transition".[37] As the mechanism has clear targets, it seeks to restore a balance by providing these regions and sectors the financial support they need to ensure a just transition. Yet, this JTM relies mostly on the idea of distributive and restorative justices which once again sheds light on the discrepancies between all nations.

6.5 Recognition Justice, International Institutions and the Just Transition

As mentioned above, recognition is logically associated with distributive justice as it recognises the rights of different groups in society.[38] Recognition has even become more important with the pandemic as it has worsened the situation of already vulnerable groups by putting jobs at risk. "While green technologies manufactories and renewable energy installations can lead to local impairments of neighbours, (measures for) climate change mitigation in general is ex- pected to have huge public health benefits, particularly for vulnerable groups".[39]

6.5.1 *The Protection of Already Vulnerable Groups*

The EIB has acknowledged that people working in informal economy are hit the hardest by the crisis in the case of Africa.[40] The example of Morocco is of particular interest. To address the needs of its vulnerable population, Morocco used mobile phone penetration "to identify and distribute financial aid to vulnerable populations using a text messaging (SMS) platform".[41] The eligible population was contacted via SMS; the government also used it to inform them about cash withdrawals points taking social distancing obligations into account. Looking after its population special needs is part of a just transition as it can be inferred from IRENA and ACE.[42] They recognise that "accelerating the deployment of renewable energy technologies is complex, as circumstances in each country differ. There is therefore no single set of solutions suited to the needs of the entire ASEAN region".[43] However, each country will consider its own situation to achieve their just transition. Interestingly, most reports have outlined themes that are redundant in most regions like women, poverty[44] and low-skill workers.

The reports have highlighted the importance of recognition in the just transition especially of vulnerable groups, but within these groups, subcategories exist like the indigenous communities, workers, consumers and women.

The indigenous communities are likely to be affected in limited countries, but within these countries, the indigenous represents may represent a big part of the population like in Canada, Australia, Brazil and Cambodia, and as a result, policies to a just transition need to consider them as well. "Equity for marginalized communities, Black, Indigenous, and people of colour (BIPOC)— especially Indigenous inclusion—was the second most cited tenet of just transition to a low- carbon economy overall" in Canada according to the International Institute for Sustainable Development (IISD).[45]

6.5.2 Women's Role in the Just Transition

Among all reports, it appears that most of the international institutions have realised the importance of supporting women during the just transition. Gender diversity is one of the keys to a just transition. For instance, "the Chilean government recognises the fundamental role women play in meeting the challenges of the energy sector and is taking action to achieve its twin goal of developing its domestic energy potential while at the same time integrating more women into the sector".[46] Chile has developed the "Energia + Mujer" programme aiming at boosting women participation in the clean energy sector. Due to this programme, Chilean energy sector is now one of the least gender diverse with a participation rate of women of just 22% and only 12% of company board directors and CEOs held by women.[47] If ACE has not recently published a report on the just transition, it has acknowledged the "women roles in Energy-Climate Nexus".[48] The illustration of this position can be found in the a project called ASEAN Climate Change and Energy Project (ACCEPT) initiated to break the silos between energy and climate sector, improving energy policies to be more climate aware.[49] With 5 women out of the 8 ACCEPT team members, it can be said that the project was quite successful in integrating women in the energy sector.[50]

6.5.3 The Growing Recognition of Consumers' Rights

Interestingly, the EU has decided to focus on the social dimension of the just transition by recognising workers' rights but also consumers' rights.[51] Moreover, the EU considers consumers' rights through the scope of health. For instance, "With regard to heating, between 50 and 125 million people in the European Union are estimated to be unable to afford proper indoor thermal comfort" as it "can also exacer- bate health inequalities. Households facing a combination of low income and inefficient homes are particularly vulnerable. There are important differences between regions".[52] In most reports, international institutions concentrate on the distribution side of just transition, but in already well-developed distribution network, countries are far ahead in their transition. The same can be observed in North America as it works on reducing consumption instead of distributing energy.[53]

6.6 Cosmopolitan Justice, International Institutions and the Just Transition

Cosmopolitan justice is less researched than the other kind of justices, but it is nonetheless an important and vital part of the just transition. This justice "stems from the belief we are all citizens of the world and so have we considered the effects beyond our borders and from a global context".[54]

6.6.1 *The Institutionalisation of Just Transition*

To successfully implement a just transition, international institutions have realised that countries needed a global framework to effectively adopt a cosmopolitan approach to the just transition. For example, the First Global Forum on Just Transition aimed to "establish a global platform to share national and international experiences on just transition in country contexts, to take stock of global research and to build momentum for action on climate change, as a result of the ILO-UNFCCC partnership".[55] Building up just transition policies will then only work by sharing knowledge and resources between all nations. Yet, most reports adopt a regional approach and scarcely mention global responsibility, but, in a way, this regional approach mimics cosmopolitan justice but applies at a smaller level.

6.6.2 *The Development of Regional Cooperation*

"The ASEAN Member States vary significantly in population, economy and energy demand. Thus, a differentiated approach to expanding renewables is necessary".[56] The differences between countries do not exclude a common approach though consideration will be given to countries' capacities. IRENA and ACE recognise that "based on their relative contributions, it is clear that to meet the target, countries near the top, such as Indonesia, Vietnam, Malaysia, and Thailand, must take action. These four countries will account for 80% of the increase in share" of renewable energy.[57]

While most reports do not directly refer to cosmopolitan justice, they consider the impacts of activities on other part of the world. For instance, the IRENA has argued, "international cooperation in energy transition should be strengthened".[58] Given the increase of the world population but also acknowledging that some remote areas have not access to electricity, it is evident that transformative changes could not happen only at country level; thus, the development of cooperation between nations would be beneficial to all as it will help develop appropriate legislation and innovation and make technology more affordable for the poorest nations.[59] Additionally, it should be said that cooperation is not only limited to the regional level and cooperation could happen between the different actors, but ultimately, it will adopt an holistic approach on the just transition.

6.6.3 *The Inclusion of All Actors in the Just Transition: The Role of the Private Sector*

For instance, even the different levels can be combined. In South America, the ILO has underlined the importance of collective actions by all actors including governments, workers and employers who also constitute key groups in the just transition as argued above.[60] "A just transition not only contributes to social justice but also increases the political feasibility of climate-related policies that are vital to curbing global GHG emissions".[61] The idea of cooperation between the different actors aims to "establish and strengthen institutional and technical capacities of subnational authorities at the regional and local levels to guide the transition, and to address the necessary changes in regional economies".[62] Moreover, according to most reports of international institutions, the private sector has become an essential part of the just transition. The role of the private sector has indeed increased to support governments in their transition just like the Inclusive Guarantee. This leading microinsurance broker has not only promoted socially inclusive insurance products but also supported West African farmers and index-based crop insurance.[63]

Appendix: International Institutions Assessed in This Chapter

The International Energy Agency (IEA)

The IEA is an autonomous intergovernmental organisation founded in 1974 in the wake of the 1973 oil crisis, under the framework of the Organisation for Economic Co-operation and Development (OECD). Initially committed to mandates related to energy security and policy cooperation, the IEA was the central international forum for energy collaboration.[64] Since its conception, the IEA has evolved to adapt to transforming global energy systems and is at the heart of global dialogue on energy, authoritative statistics, analysis and the examination of the full spectrum of energy issues.[1] Moreover, policy advocation with the hope of enhancing the reliability, affordability and sustainability of energy within its member countries and beyond is needed.[1]

The IEA aids governments, industries and decision-makers to accelerate clean and just energy transitions. Further, in early 2021 the new global commission led by the Danish Prime Minister put forward "Our Inclusive Energy Future" which will examine the social and economic impacts, on individuals and communities, of cleaner energy systems.[65] The IEA's most notable report is "World Energy Outlook", providing information on how global energy systems will evolve in the coming years.[66] However, there is no direct reference to the Just Transition but rather the clean energy transition.

International Renewable Energy Agency (IRENA)

IRENA was established in 2009 and is an intergovernmental organisation aiding countries through their transition towards a sustainable energy future, serving as the primary platform for international cooperation as well as technology-, policy- and finance-based knowledge on renewable energy.[67] IRENA supports governments to enable them to adopt renewable energy policies and investments.

IRENA has recently proposed a "World Energy Transitions Outlook" outlining international strategies towards carbon-neutrality as well as a climate-safe 1.5°C pathway by 2050.[68] More specifically, the outlook will address concerns about the just transition through calls for a comprehensive and consistent policy framework.[5] The report has a chapter dedicated to "Broad, Holistic Just Transition Policies" which emphasises the essentiality for policy framework to achieve targets effectively.[5] Moreover, it outlines that policies which support and enable the deployment of renewable energy must be in conjunction with policies that ensure industrial and other economic capabilities that will foster a just transition through social protection, labour, skills and education.[5]

United Nations (UN)

The United Nations is an international, intergovernmental organisation established in 1945. The primary aims of the UN are to maintain peace and security around the world while developing relations between nations and achieving international cooperation.[69] Member states now include 193 nations who are all members of the General Assembly. Notable bodies of work produced by the UN include the Charter of the United Nations, the Universal Declaration of Human Rights and the Statute of the International Court of Justice.[6]

As part of the global acknowledgement for immediate climate action, the UN created the Sustainable Development Goals (SDGs). The SDGs are part of the 2030 Agenda for Sustainable Development adopted by all members states in 2015.[70] Thus far, there are 169 targets and there have been 3036 events, 1257 publications and 5414 actions.[7] Goals relating to the just transition include goal 8, decent work and economic growth, and goal 13, climate action. Additionally, in 2019, the UN Secretary-General hosted the Climate Action Summit where a clear direction for climate action was set, but there was no direct mention of the Just Transition.[71]

World Bank

The World Bank, established in 1944, is an international finance institution providing loans and grants to governments in low-middle-income countries around the world.[72] The group has a global partnership with five institutions working towards sustainable solutions to reduce poverty and build prosperity in developing countries.[9]

Recently, the World Bank published a brief: "Coal Mine Closure and a Just Transition for All". The brief highlights how the closure of mines will impact coal-dependent communities, for example, and has put the World Bank at the centre of global discussions regarding the energy transition.[73] An example of such an initiative was in early 2021 when the World Bank, in collaboration with the European Commission, pledged to assist Poland in order to ensure a just transition in the mining regions within the country.[74]

International Monetary Fund (IMF)

The IMF, like the World Bank, is an international financial institution and was established in 1944, in the wake of the Great Depression.[75] The founding members strived to create a framework for international economic cooperation; as of today, there are 190 member countries.[12] The fund surveys the international monetary system and global economic developments to identify risks to then recommend policies for financial stability and growth.[12]

The IMF's managing director made a speech in April 2021 on "securing a green recovery: the economic benefits from tackling climate change". While the speech did not directly mention the just transition, there was reference to securing a green recovery through climate-resilient economies, climate pricing to curb emissions and green financing.[76]

Asian Development Bank

ADB was founded in 1966 and is a regional development bank headquartered in the Philippines. The ADB's focus is primarily in Asia envisioning a prosperous, inclusive, resilient and sustainable Asia and Pacific region.[77]

In 2018 ADB hosted a seminar titled: "High-level Dialogue on More Quality Jobs in Asia and the Pacific: Investing for a Just Transition".[78] The seminar was in collaboration with the International Labour Organisation and the Government of Philippines and was heavily centred around the facilitation and implementation of the just transition to meet with Paris Agreement commitments.[15] Additionally, in 2020 they produced a report on green finance strategies in a post-COVID-19 world. The report does not make direct reference to a just transition but rather promotes partnerships to build a resilient, green and socially just recovery in Asia.[79]

European Investment Bank (EIB)

The EIB is the European Union's (EU) investment bank founded in 1958 and owned by EU member states. EIB is the largest multilateral financial institution in the world and is also one of the largest suppliers of climate finance.[80]

The EIB and the European Commission are working in conjunction towards a Just Transition Mechanism that proposes a public loan facility that will finance green investment in the EU.[81] Further, in late 2020 the EIB produced the EIB group climate bank roadmap 2021–2025, which mentioned accelerating the transition through green finance as well as ensuring a just transition for all.[82]

Caribbean Development Bank

The CDB was established in Jamaica, 1969, and put into force in 1970. CDB is a regional financial institution whose ethos incorporates harmonious economic growth and development in the Caribbean.[83] The CDB strives to halve extreme poverty in the region by 2025 through sustainable growth and good governance.[20]

One sector of the CDB is "energy generation, distribution and efficiency", where the focus is to improve energy access and the acceleration of a transition towards sustainable energy.[84] While there is no direct mention of the just transition, the bank has released an energy transition document titled "jump-starting the Caribbean's energy transition: battery storage and grid modernisation",[85] which is a positive step towards adopting just transition policies.

International Labour Organisation

The ILO is a UN agency comprised of 187 member states and it was established in 1919.[86] The primary goals of the ILO are to set standards of labour and policy development and to generate programmes that promote decent work for all. The ILO is unique in its tripartite structure which allows all workers, employers and governments to have an equal voice,[23] thus, ensuring that social partners are reflected in labour standards and through policy and programme shaping.[23]

In 2015 the ILO produced a report titled, "Guidelines for a just transition towards environmentally sustainable economies and societies for all".[87] The report details how the ILO will work to fulfil the goals outlined in the Paris Agreement and how the just transition can work practically as well as how trade unions and workers can contribute.[24] Moreover, in 2017 they released a report, "A Just Transition to a Sustainable Future—Next steps for Europe".[88] The report details creating green and decent jobs and a concise list of the ILO principles for a just transition.[25]

ASEAN Centre for Energy (ACE)

ACE is an intergovernmental organisation within the Association of Southeast Asian Nations (ASEAN) which represents ten ASEAN member states' interests within the energy sector.[89] Established in 1999, ACE acts as a catalyst for integration and economic growth in ASEAN countries by providing expertise and information necessary to inform energy policies and programmes that coincide with regional environmental sustainability.[26] ACE operates under the outputs of three roles: catalysts, knowledge hubs and think tanks.[26]

In March of 2021 ACE hosted a webinar discussing the empowerment of women of ASEAN as a key player of the just energy transition.[90] The webinar itself brought into question gender issues related to the just energy transition with an emphasis of the role women will play in both energy and climate sectors.[27]

European Union

The EU comprises 27 member states and refers to the political and economic union of all member states.[91] Headquartered in Brussels, the main goals of the EU are to promote peace and offer freedom, security and justice as well as sustainable development. Moreover, sustainable development relies on balanced economic growth, price stability, employment and environmental protection.

The European Commission has recently welcomed a political agreement by the European Parliament and Council on a proposal for a Just Transition Fund (JTF).[92] With a budget of €17.5 billion, the fund is integral to the European Green Deal and the first pillar of the Just Transition Mechanism.[29] The fund emphasises social and economic stability through diversifying economic activity and aiding workers towards a changing labour market.[29]

The Just Transition Mechanism is a supportive framework which promotes just transition efforts. It will provide financial resources, including the €17.5bn JTF and €13.5bn in grants and loans, and technical assistance to the EU.[93]

Commonwealth

The Commonwealth (CW) refers to a voluntary political association formed of 54 member states all of whom were once territories of the British Empire.[94] The CW comprises three intergovernmental organisations: The Commonwealth Secretariat, The Commonwealth Foundation and the Commonwealth of Learning.

The Commonwealth has produced a Sustainable Energy Transition Agenda (CSET) that aids in the acceleration of action by member countries to achieve SDG 7, affordable and clean energy. Three pillars have been implemented to achieve this: inclusive transitions, technology and innovation and enabling frameworks.[95] Moreover, the CW has acknowledged that the energy transition will require strong political ambition from member states in order to establish new frameworks. Additionally, they have highlighted that the energy transition can leave no one behind.[32]

International Institute for Sustainable Development (IISD)

The IISD is an independent think tank founded in 1990 with offices across Canada. Their mission is to create solutions to ensure a stable climate with sustainable resources and fair economies.[96] IISD receives the majority of their funding from governments inside and outside of Canada, UN agencies, foundations and the private sector.[33]

In 2018 the IISD published a report, "Real People, Real Change: Strategies for Just Energy Transitions", aiming to aid governments to make their energy transitions just.[97] The report is based on case study research from Canada, Egypt, Indonesia, India, Poland and Ukraine.[34] Additionally, in 2019 the IISD hosted an event regarding strategies related to the just transition with reference to shifting to below 2°C economies.[98] Finally, another report was published titled: "Fossil Fuel Subsidy Reform and the Just Transition: Integrating approaches for complementary outcomes".[99]

Greenpeace

Greenpeace was first established in 1971 by environmental activists and is a non-governmental environmental organisation.[100] Their efforts rely solely on public donations, and they do not accept funding from governments, corporations or political parties.[31] This allows them to confront governments and corporations who are responsible for destroying the natural environment to affect real change.[31]

Greenpeace performed research on how the energy transition would impact oil and gas workers. They generated a report, which gives a voice to workers in industries set to be phased out and highlights their perspectives and priorities.[101] Their website also features a blog post outlining the parameters surrounding the just transition such as job opportunities in a clean economy, the essentiality for a just transition and a case study illustrating a successful energy transition.[102] Additionally, Greenpeace have published a "Green Recovery" manifesto, which references COVID-19 as a huge challenge to the climate crisis as well as varying recovery packages as well as funding and investment proposals.[103]

Friends of the Earth (FoE)

FoE is an environmental campaigning community comprised of lawyers, local groups and supporters fighting for environmental and social justice around the world. FoE strive to reduce plastic waste, save EU environmental protections, beat climate breakdown, protect nature and double tree cover.

FoE Scotland has generated a ten-step plan to green job recovery which calls for government intervention to create good jobs that will deliver a just and green recovery.[104] FoE has also complied a climate action plan which highlights the six key points that it believes will help the UK government to tackle climate change: transport, power, buildings and homes, trees and food, consumption and international justice.[105] While it does not directly mention the just transition, the plan coincides with many of the just transition goals.

Notes

1. Alan M Rugman and Jonathan P Doh, 'The Role of International Institutions', *Multinationals and Development* (Yale University Press 2017) 103.
2. International Renewable Energy Agency and ASEAN Centre for Energy, 'Renewable Energy Outlook for ASEAN: a REmap Analysis' (IRENA, 2016) 18. https://www.irena.org/-/media/Files/IRENA/Agency/Publication/2016/IRENA_REmap_ASEAN_2016_report.pdf (last accessed 1 September 2021).
3. Ibid., 75.
4. European Investment Bank, 'The rise of Africa's digital economy—The European Investment Bank's activities to support Africa's transition to a digital economy' (EIB Staff paper, February 2021) 47. https://www.eib.org/attachments/thematic/study_the_rise_of_africa_s_digital_economy_en.pdf (last accessed 1 September 2021).
5. Ibid.
6. Ibid.
7. Ibid., 65.
8. Ibid.

9. Ibid.
10. RGA, 'Back to Basics? Evolution of Distribution Models in Inclusive (Micro) Insurance' (27 August 2020). https://www.rgare.com/knowledge-center/media/articles/back-to-basics-evolution-of-distribution-models-in-inclusive-(micro)-insurance (last accessed 1 September 2021).
11. Ibid.
12. Raphael J Heffron and others, 'How Different Electricity Pricing Systems Affect the Energy Trilemma: Assessing Indonesia's Electricity Market Transition' (Asian Development Bank Institute, 2021) 19. https://www.adb.org/sites/default/files/publication/674226/adbi-wp1213.pdf (last accessed 1 September 2021).
13. Ibid., 22.
14. Serena Olsaretti, *The Oxford Handbook of Distributive Justice* (OUP 2018) 2.
15. IRENA and ACE, 'Renewable Energy Outlook for ASEAN: a REmap Analysis' (IRENA, 2016) 40. https://www.irena.org/-/media/Files/IRENA/Agency/Publication/2016/IRENA_REmap_ASEAN_2016_report.pdf (last accessed 1 September 2021).
16. Ibid., 63.
17. Raphael J Heffron and others, 'How Different Electricity Pricing Systems Affect the Energy Trilemma: Assessing Indonesia's Electricity Market Transition' (Asian Development Bank Institute, 2021) 22. https://www.adb.org/sites/default/files/publication/674226/adbi-wp1213.pdf (last accessed 1 September 2021).
18. IRENA, '*World Energy Transitions Outlook: 1.5°C Pathway*' (IRENA, 2021) 49. https://www.irena.org/-/media/Files/IRENA/Agency/Publication/2021/March/IRENA_World_Energy_Transitions_Outlook_2021.pdf (last accessed 1 September 2021).
19. EIB, 'The rise of Africa's digital economy—The European Investment Bank's activities to support Africa's transition to a digital economy' (EIB Staff paper, February 2021) 27. https://www.eib.org/attachments/thematic/study_the_rise_of_africa_s_digital_economy_en.pdf (last accessed 1 September 2021).
20. Ibid.
21. Ibid.
22. Ibid., 41.
23. Catherine Saget, Adrien Vogt-Schilb and Trang Luu, 'Jobs in a Net-Zero Emissions Future in Latin America and the Caribbean' (Inter-American Development Bank and International Labour Organization, 2020) 91. https://www.ilo.org/wcmsp5/groups/public/%2D%2D-

americas/%2D%2D-ro-lima/documents/publication/wcms_752069.pdf (last accessed 1 September 2021).
24. Ibid., 69.
25. Ibid.
26. Ibid., 89.
27. Raphael J Heffron, 'The role of justice in developing critical minerals' 2020 7(3) The Extractive Industries and Society 855–863, 858. And see also: Hazrati, M. and Heffron, R. J. 2021. Conceptualising restorative justice in the energy transition: changing the perspectives of fossil fuels. Energy Research and Social Science, 78, 102115.
28. EIB, 'The rise of Africa's digital economy—The European Investment Bank's activities to support Africa's transition to a digital economy' (EIB Staff paper, February 2021) 77. https://www.eib.org/attachments/thematic/study_the_rise_of_africa_s_digital_economy_en.pdf (last accessed 1 September 2021).
29. Ibid.
30. Catherine Saget, Adrien Vogt-Schilb and Trang Luu, 'Jobs in a Net-Zero Emissions Future in Latin America and the Caribbean' (Inter-American Development Bank and International Labour Organization, 2020) 102. https://www.ilo.org/wcmsp5/groups/public/%2D%2D-americas/%2D%2D-ro-lima/documents/publication/wcms_752069.pdf (last accessed 1 September 2021).
31. Ibid.
32. Ibid.
33. Ibid., 76.
34. African Development Bank Group, 'Green Growth In Africa' (AfDB, 2014) 5. https://www.afdb.org/fileadmin/uploads/afdb/Documents/Project-and-Operations/Supporting_Africa's_Transition_to_Green_Growth_-_Snapshot_of_the_AfDB's_Activities.pdf (last accessed 1 September 2021).
35. Geothermal Development Company, 'Menengai Geothermal Project (*GDC*, 2020) 102. https://www.gdc.co.ke/menengai.html (last accessed 1 September 2021).
36. Dirk Arne Heyen, Luisa Menzemer, Franziska Wolff, Andreea Beznea and Rob Williams, 'Just transition in the context of EU environmental policy and the European Green Deal' (Öko-Institut, 2020) 9. https://ec.europa.eu/environment/enveco/growth_jobs_social/pdf/studies/just_transition_issue_paper_final_clean.pdf (last accessed 1 September 2021).
37. ibid.
38. Raphael J Heffron, 'The role of justice in developing critical minerals' 2020 7(3) The Extractive Industries and Society 855–863, 858.

39. Arumdari Nurgianti and Zulfikar Yurnaidi, 'Women Roles in Energy-Climate Nexus' (8 March 2020). https://accept.aseanenergy.org/women-roles-in-energy-climate-nexus/ (last accessed 1 September 2021).
40. EIB, 'The rise of Africa's digital economy—The European Investment Bank's activities to support Africa's transition to a digital economy' (EIB Staff paper, February 2021) 43. https://www.eib.org/attachments/thematic/study_the_rise_of_africa_s_digital_economy_en.pdf (last accessed 1 September 2021).
41. Ibid.
42. Ibid.
43. IRENA and ACE, 'Renewable Energy Outlook for ASEAN: a REmap Analysis' (IRENA, 2016) 79. https://www.irena.org/-/media/Files/IRENA/Agency/Publication/2016/IRENA_REmap_ASEAN_2016_report.pdf (last accessed 1 September 2021).
44. Olivier De Schutter, 'Report on the "just transition" in the economic recovery: eradicating poverty within planetary boundaries' (OHCHR, 2020). https://www.ohchr.org/en/Issues/Poverty/Pages/CallforsubmissionsJustTransition.aspx (last accessed 1 September 2021).
45. Estan Beedell and Vanessa Corkal, 'Building Momentum for a Just Transition in Canada: Perspectives from civil society' (IISD, 2021) 6. https://www.iisd.org/system/files/2021-04/building-just-transition-canada-civil-society.pdf (last accessed 1 September 2021).
46. IEA, 'Chile's "Energia + Mujer" programme seeks to boost participation of women in the clean energy sector' (30 October 2018). https://www.iea.org/news/chiles-energia-mujer-programme-seeks-to-boost-participation-of-women-in-the-clean-energy-sector (last accessed 1 September 2021).
47. Ibid.
48. Arumdari Nurgianti and Zulfikar Yurnaidi, 'Women Roles in Energy-Climate Nexus' (8 March 2020). https://accept.aseanenergy.org/women-roles-in-energy-climate-nexus/ (last accessed 1 September 2021).
49. Ibid.
50. Ibid.
51. Dirk Arne Heyen, Luisa Menzemer, Franziska Wolff, Andreea Beznea and Rob Williams, 'Just transition in the context of EU environmental policy and the European Green Deal' (Öko-Institut, 2020) 12;19. https://ec.europa.eu/environment/enveco/growth_jobs_social/pdf/studies/just_transition_issue_paper_final_clean.pdf (last accessed 1 September 2021).
52. Ibid., 18.
53. Samantha Smith, 'Just Transition: A Report for the OECD' (Just Transition Centre, 2017) 10. https://www.oecd.org/environment/cc/

g20-climate/collapsecontents/Just-Transition-Centre-report-just-transition.pdf (last accessed 1 September 2021).
54. Raphael J Heffron, 'The role of justice in developing critical minerals' 2020 7(3) The Extractive Industries and Society 855–863, 858.
55. ILO, 'Final Report of the first Global Forum on Just Transition ' (ILO and the UN Climate Change secretariat (UNFCCC), 2017) 3. https://www.un-page.org/files/public/global_forum_on_just_transition_final_report3.pdf (last accessed 1 September 2021).
56. IRENA and ACE, 'Renewable Energy Outlook for ASEAN: a REmap Analysis' (IRENA, 2016) 47. https://www.irena.org/-/media/Files/IRENA/Agency/Publication/2016/IRENA_REmap_ASEAN_2016_report.pdf (last accessed 1 September 2021).
57. Ibid.
58. IRENA, '*World Energy Transitions Outlook: 1.5°C Pathway*' (IRENA, 2021) 50. https://www.irena.org/-/media/Files/IRENA/Agency/Publication/2021/March/IRENA_World_Energy_Transitions_Outlook_2021.pdf (last accessed 1 September 2021).
59. Ibid.
60. Catherine Saget, Adrien Vogt-Schilb and Trang Luu, 'Jobs in a Net-Zero Emissions Future in Latin America and the Caribbean' (Inter-American Development Bank and International Labour Organization, 2020) 92. https://www.ilo.org/wcmsp5/groups/public/%2D%2D-americas/%2D%2D-ro-lima/documents/publication/wcms_752069.pdf (last accessed 1 September 2021).
61. Ibid.
62. ILO, 'Guidelines for a just transition towards environmentally sustainable economies and societies for all' (Inter-American Development Bank and International Labour Organization, 2015) 8. https://www.ilo.org/wcmsp5/groups/public/%2D%2D-ed_emp/%2D%2D-emp_ent/documents/publication/wcms_432859.pdf (last accessed 1 September 2021).
63. European Investment Bank, 'The rise of Africa's digital economy—The European Investment Bank's activities to support Africa's transition to a digital economy' (EIB Staff paper, February 2021) 47. https://www.eib.org/attachments/thematic/study_the_rise_of_africa_s_digital_economy_en.pdf (last accessed 1 September 2021).
64. IEA. *History—About—From oil security to steering the world toward secure and sustainable energy transitions.* (2021). https://www.iea.org/about/history (last accessed 1 September 2021).
65. IEA. *Our Inclusive Energy Future—Programmes—IEA*. (2021). https://www.iea.org/programmes/our-inclusive-energy-future (last accessed 1 September 2021).

66. IEA. *World Energy Outlook 2020*, IEA, Paris. (2020). https://www.iea.org/reports/world-energy-outlook-2020 (last accessed 1 September 2021).
67. Irena.org. *About IRENA*. (2021). https://www.irena.org/aboutirena (last accessed 1 September 2021).
68. IRENA. *World Energy Transitions Outlook, 1.5C Pathway—Preview*. (2021). https://irena.org/publications/2021/March/World-Energy-Transitions-Outlook (last accessed 1 September 2021).
69. United Nations. *About Us | United Nations*. https://www.un.org/en/about-us (last accessed 1 September 2021).
70. United Nations Department of economic and Social Affairs -Sustainable Development. *THE 17 GOALS | Sustainable Development*. https://sdgs.un.org/goals (last accessed 1 September 2021).
71. United Nations—Climate Action. *2019 Climate Action Summit*. (2019). https://www.un.org/en/climatechange/2019-climate-action-summit (last accessed 1 September 2021).
72. World Bank. *Who We Are*. https://www.worldbank.org/en/who-we-are (last accessed 1 September 2021).
73. World Bank. *Coal Mine Closure and a Just Transition for All*. (2020). https://www.worldbank.org/en/results/2020/11/10/coal-mine-closure-and-a-just-transition-for-all (last accessed 1 September 2021).
74. World Bank. *World Bank and the European Commission to Support Poland to Transition out of Coal*. (2021). https://www.worldbank.org/en/results/2020/11/10/coal-mine-closure-and-a-just-transition-for-all (last accessed 1 September 2021).
75. International Monetary Fund. *About the IMF—We are a global organisation, surveillance*. (2021). https://www.imf.org/en/About (last accessed 1 September 2021).
76. Georgieva, K. *Securing a Green Recovery: The Economic Benefits from Tackling Climate Change*. (2021). https://www.imf.org/en/News/Articles/2021/04/15/sp041521-securing-a-green-recovery (last accessed 1 September 2021).
77. Asian Development Bank. *Who we are—About ABD*. https://www.adb.org/who-we-are/about (last accessed 1 September 2021).
78. GoP, ADB, ILO. *High-level Dialogue on More Quality Jobs in Asia and the Pacific: Investing for a Just Transition*. (2018). https://www.adb.org/annual-meeting/2018/events/quality-jobs (last accessed 1 September 2021).
79. ADB. *Green Finance Strategies for Post-COVID19 Economic Recovery in Southeast Asia*. (2020). https://www.adb.org/publications/green-finance-post-covid-19-economic-recovery-southeast-asia (last accessed 1 September 2021).

80. EIB. *Who we are.* (2021). https://www.eib.org/en/about/index.htm# (last accessed 1 September 2021).
81. EIB. *Just Transition Mechanism: the EIB and the European Commission join forces in a proposed new pubic loan facility to finance green investments in the EU.* (2020). https://www.eib.org/en/press/all/2020-130-commission-proposes-a-public-loan-facility-to-support-green-investments-together-with-the-eib (last accessed 1 September 2021).
82. EIB. *EIB Group Climate Bank Roadmap 2021–2025.* (2020). https://www.eib.org/en/publications/the-eib-group-climate-bank-roadmap (last accessed 1 September 2021).
83. CDB. *Bank History—Overview.* https://www.caribank.org/bank-history (last accessed 1 September 2021).
84. CDB. *Sectors—Energy generation, distribution and efficiency.* https://www.caribank.org/our-work/sectors/energy-generation-distribution-and-efficiency (last accessed 1 September 2021).
85. CDB. *Jump-starting the Caribbean's energy transition: battery storage and grid modernisation.* (2017). https://www.caribank.org/publications-and-resources/resource-library/booklets-brochures/jumpstarting-caribbeans-energy-transition-battery-storage-and-grid-modernisation (last accessed 1 September 2021).
86. ILO. *About the ILO.* https://www.ilo.org/global/about-the-ilo/lang%2D%2Den/index.htm (last accessed 1 September 2021).
87. ILO. *Guidelines for a just transition towards environmentally sustainable economies and societies for all.* (2015). https://www.ilo.org/wcmsp5/groups/public/%2D%2D-ed_emp/%2D%2D-emp_ent/documents/publication/wcms_432859.pdf (last accessed 1 September 2021).
88. ILO. *A Just Transition to a Sustainable Future—Next steps for Europe.* (2017). https://www.ilo.org/wcmsp5/groups/public/%2D%2D-europe/%2D%2D-ro-geneva/%2D%2D-ilo-brussels/documents/publication/wcms_614024.pdf (last accessed 1 September 2021).
89. ASEAN Centre for Energy. *Introductions- Vision and Mission ACE.* https://aseanenergy.org/introductions/ (last accessed 1 September 2021).
90. ASEAN Centre for Energy. *Empowering Women of ASEAN as the Key Player of Just Energy Transition.* (2021). https://aseanenergy.org/event/empowering-women-of-asean-as-the-key-player-of-just-energy-transition/ (last accessed 1 September 2021).
91. European Union. *The EU in brief—Goals and Values of the EU.* https://europa.eu/european-union/about-eu/eu-in-brief_en (last accessed 1 September 2021).
92. European Commission. *European Commission welcomes the political agreement on the Just Transition Fund. Press release* (2020). https://ec.

europa.eu/commission/presscorner/detail/en/IP_20_2354 (last accessed 1 September 2021).

93. World Resources Institute—Just Transition and Equitable Climate Action. *European Union's Just Transition Mechanism: Transnational Funding and Support for a Just Transition.* (2020). https://www.wri.org/just-transitions/european-union (last accessed 1 September 2021).
94. The Commonwealth. *About us.* https://thecommonwealth.org/about-us (last accessed 1 September 2021).
95. The Commonwealth. *Commonwealth Sustainable Energy Transitions.* (2021). https://thecommonwealth.org/commonwealth-sustainable-energy-transitions-agenda-children's-books-energy-transition (last accessed 1 September 2021).
96. IISD. *Mission and Goals.* https://www.iisd.org/mission-and-goals (last accessed 1 September 2021).
97. Zinecker. A, Gass. P, Gerasimchuk. I et al. *Real People, Real Change, Strategies for just energy transitions.* (2018). https://www.iisd.org/publications/real-people-real-change-strategies-just-energy-transitions (last accessed 1 September 2021).
98. Gaas, P. *In Search of the Just Transition: Examples from around the world-Further reading.* (2019). https://www.iisd.org/articles/just-transition-examples (last accessed 1 September 2021).
99. Gaas. P, Echeverria. *Fossil Fuel Subsidy Reform and the Just Transition.* (2017). https://www.iisd.org/publications/fossil-fuel-subsidy-reform-and-just-transition-integrating-approaches-complementary (last accessed 1 September 2021).
100. Greenpeace. *About Greenpeace.* https://www.greenpeace.org.uk/about-greenpeace/ (last accessed 1 September 2021).
101. Greenpeace. *Offshore: Oil and Gas Worker' Views on Industry Conditions and the Energy Transition.* (2020). https://www.greenpeace.org.uk/resources/offshore-oil-and-gas-workers-report/ (last accessed 1 September 2021).
102. Greenpeace. *The Just Transition.* https://www.greenpeace.org.uk/challenges/just-transition/ (last accessed 1 September 2021).
103. Greenpeace. *Green Recovery Manifesto.* (2020). https://www.greenpeace.org.uk/resources/green-recovery-manifesto/ (last accessed 1 September 2021).
104. FoE Scotland. *10 Steps to a Green Jobs Recovery.* (2020). https://foe.scot/10-steps-green-jobs-recovery/ (last accessed 1 September 2021).
105. FoE. *How the UK government can tackle climate breakdown, Our 6-point Climate Action Plan.* (2020). https://friendsoftheearth.uk/climate/how-uk-government-can-tackle-climate-breakdown-our-6-point-climate-action-plan (last accessed 1 September 2021).

CHAPTER 7

Conclusion: The Just Transition Movement Post Pandemic, COP26 and the Financial Crisis

Abstract In this chapter, the future of the just transition movement is considered in light of COP26, COVID-19 and the current financial crisis. These are all challenges in themselves; however, despite these society continues its desire to have a just transition to a low-carbon economy. And certainly, current law and policy development in the just transition demonstrates that it is a movement that is set to continue. Also briefly explored is the research gap in the literature around financing the just transition. Finally, re-highlighted and summarised is the importance of the role of law and how it is central to achieve the just transition to a low-carbon economy.

Keywords COP26 • Just transition • Financial crisis • COVID-19 • Financing the just transition

7.1 Introduction: The Just Transition Movement Continues

The COVID-19 pandemic continues, and it is certain that its legacy will be a continued financial crisis. In this context, there have already been significant calls for a green economic recovery by the majority of international institutions. Further, within this same time period, the UN Secretary-General has called for all countries to declare a climate

R. J. Heffron, *Achieving a Just Transition to a Low-Carbon Economy*, https://doi.org/10.1007/978-3-030-89460-3_7

emergency in 2020.[1] Despite all these issues, the debate around moving to a just transition to a low-carbon economy continues.

The development of policy initiatives to achieve a just transition in society is growing across different countries worldwide. It is clear that the just transition has moved beyond its origin in exploring issues of around labour, and it now has the far broader goal of ensuring that all society moves to a just way to a low-carbon economy.

A key issue is how can the just transition be delivered post pandemic and within the context of the economic crisis—this will be referred to as the post-pandemic world. Previous plans of how to achieve a just transition have in 2019 and early 2020 have been impacted by the pandemic. In particular, governments have had to readjust their budgets. There have been some positives, however, as it could be argued that that plans pre-pandemic were insufficient and slow. The pandemic has provided an opportunity to accelerate some of the policy actions necessary to meet a just transition.

7.2 The Development of National Commissions and Subsequent Financing for a Just Transition

As stated previously, the just transition policy journey from a legal perspective began really in 2015 when it received a mention in the 2015 Paris Agreement. It was discussed prior to this in the international policy arena (in particular, from a jobs perspective), but 2015 really marked its progress in international and national legal terms. The 2015 Paris Agreement is also significant in that never before have so many countries (188) signed and ratified an international agreement so fast.[2]

It took several years for the policy community to realise that the issue of a "just transition" was mentioned within the aforementioned international legal treaty.[3] It received impetus in 2018 when at the G7 talks the issue of a just transition was specifically mentioned as a policy goal in the final communique issued by the governments.[4]

In 2019, some countries began to deliver and develop just transition policies at a national level, and these are detailed in this book which is the first to make a legal contribution on the just transition to a low-carbon economy. As detailed some countries put in place the first legislative steps to achieve a just transition through the creation of what can be classified as a *Just Transition Commission (JTC)*. There are a number of different forms, but they all involve the establishment of some form of commission. This text explores these issues by looking at a range of different countries, including Canada, Germany, Scotland, Australia, Ireland, New Zealand, the US (Appalachia), South Africa and the EU.

In 2020, despite the emergence of the pandemic, the key change in terms of just transition policy has been the discussion and also planned allocation of funding plans for the just transition. This has in essence built on the 2019 legal developments and specifically has begun in the EU, Germany and Scotland in 2020 and continues in 2021. One of the big areas of scholarship will be how to finance the transition, and this has already been researched by a great team of academics based at the London School of Economics, and they continue to do great work in the area. However, it is an under-researched area, and there will be other researchers who move into this space over the next few years.

7.3 Further Next Steps: Implementing the Just Transition

Not all countries are creating these JTCs, but they are utilising other methods to achieve a just transition. In essence they involve a similar set of actions but may take longer to materialise as they are not all coordinated under the same "umbrella" action of a JTC. These actions involve utilising existing legal frameworks which are mainly:

1. National and regional development plans
2. Strategic environmental assessments
3. National industrial strategies

Collectively actions under these three interrelated frameworks can also enable the implementation of a just transition. It is certain that national and regional development plans and national industrial strategies will focus on job creation, while strategic environmental assessments will cover energy, environmental and climate policy issues. All three legal frameworks will have justice as an additional outcome to ensure a just transition.

Significantly, for policy-makers, these three steps may involve less opposition since they are existing frameworks in many countries already. A JTC may take several years to establish and another few years to deliver its results. Further the effectiveness of its results will be dependent on the powers of the JTC. In contrast, amendments can begin on national and regional development plans, strategic environmental assessments and national industrial strategies can begin immediately, and hence these frameworks may deliver a just transition faster than a JTC. These amendments, however, all need to have outcomes of a just transition such as more fairness, equality, equity and inclusiveness at their core.

7.3.1 *A Role for Data*

Today, society is characterised by data. It is no surprise therefore that data on the energy sector is on rise. The availability of this data, the transferability of this data and the accuracy of this data is informing energy decision-making. In policy analysis, in legal courtrooms and in company boardrooms, there is a need to accurately assess the data as it is becoming easier to demonstrate the socio-economic, environmental and climate change and general societal (inc. health impacts) impacts of an energy project. Accountability in decision-making will increase as it will be harder to avoid doing more comprehensive data analysis.

A just energy transition will need all stakeholders to work together, and that includes those who are overly reliant on FF today. For the research community, there is an opportunity and need to work collaboratively like never before. To deliver just outcomes post this pandemic and given the climate change emergency, more innovative and interdisciplinary methods of ensuring society meets its energy and climate targets of 2030 are necessary. Data will form the cornerstone of much of the interdisciplinary and collaborative research, and strategically it can be utilised to deliver a just energy transition.

7.3.2 *Post COP26*

It is my anticipation that human rights in the energy sector and then more broadly within the just transition to a low-carbon economy will be an area that will increase in professional and research practice post COP26. Already this is evident as highlighted in terms of challenges to multinational energy companies.[5] The protection of these human rights through court rooms will add a more systematic approach to ensuring a just energy transition for 2030 under the 2015 Paris Agreement and also in terms of net zero targets.[6]

Policy-makers will become accountable for how they meet their commitments under the 2015 Paris Agreement. They will have to account for how they have adjusted, revised or added to their plans for COP26 and then later again when it is required for 2025 at COP30. Both these time periods increase the likelihood of a wave of legal action, as different stakeholders take action on governments or other key stakeholders and seek out more demanding energy and climate goals for COP26 and then subsequently for COP30. And at the heart of this legal action will be how the policy-makers aim to ensure the entire list of human rights were protected within these energy transition and climate plans.

The evidence to support the continuation of fossil fuel investments is becoming harder to obtain and also harder to demonstrate successfully. Already the savvy investors must be looking at the proposed bans on coal and fossil fuel-based cars and planning exit strategies for the next fossil fuel activities that are under threat as they would be valued at a peak price at the moment (and therefore it is the opportune time to sell).

7.3.3 Timelines and the Role of Law in the Just Transition

Finally, it is necessary to focus on the role of timelines. If countries are to solve the climate emergency issue and ensure that they meet the 2015 Paris Agreement energy and climate targets of 2030, it is vital that institutions, stakeholders, plans and policies are all established or updated. Law plays its role here to ensure that the new roles of institutions and stakeholders are defined (or redefined). Then all plans and policies have to be legislated for and put into general legislation. For example, to meet 2030 energy and climate targets, action needs to happen now in terms of creating new legislation and/or institutions; they will take several years to be established and more to be effective and deliver for policy objectives for 2030. This is the advantage of utilising the *Just Framework* in that it ensures a scholar engages with the timelines for action and ensures realistic policy goals that also just.

A vital role of law in the just transition is that it can stimulate the investment needed to ensure a just transition to a low-carbon economy. Investors need legal certainty in their investment, that is, they need to be assured that they will have sufficient time to recoup a return on their investment. An investor will look at how a country sets out to achieve a just transition to a low-carbon economy. They will explore how institutions and policies are aligned and therefore whether a country represents a good investment from a just transition perspective; this type of analysis was provided in late 2018 by one major multinational bank for its investors and how they should consider developing countries and their moves towards a just transition.[7]

The key issue is the importance of the role of law in achieving a just transition to a low-carbon economy. This is particularly important post pandemic when governments worldwide aim to kick-start their economies out of the crisis. Major public investment programmes are planned, and private sector investment is being stimulated with an ongoing stream of investments. In these economic recovery trajectories, no one should be left behind, and therefore the just transition needs to be at the heart of societies' post pandemic policy action.

Legislating for the just transition is vital to secure this investment whether it is through creating JTCs or widespread amendments to national and regional development plans, strategic environmental assessments and national industrial strategies. In considering the climate emergency—consider the flash floods this year in 2021 in Germany, China and the UK being attributed to climate change—coupled with the economic crisis and pandemic, *there is a need for the just transition evolution to become more revolutionary, and the aim of this book is to demonstrate that law can be at the centre of achieving that.*

Notes

1. United Nations. 2020 (12 Dec). *Secretary-General's remarks at the Climate Ambition Summit.* Available at: https://www.un.org/sg/en/content/sg/statement/2020-12-12/secretary-generals-remarks-the-climate-ambition-summit-bilingual-delivered-scroll-down-for-all-english-version (last accessed 1 September 2021).
2. UN. 2020a. *2015 Paris Agreement.* Available at: https://unfccc.int/process/the-paris-agreement/status-of-ratification (last accessed 1 September 2021).
3. UN. 2020b. *2015 Paris Agreement.* Available at: https://unfccc.int/files/essential_background/convention/application/pdf/english_paris_agreement.pdf (last accessed 1 September 2021).
4. European Council. 2018. *The Charlevoix G7 Summit Communique.* Available at: https://www.consilium.europa.eu/en/press/press-releases/2018/06/09/the-charlevoix-g7-summit-communique/ (last accessed, 1 September 2021).

5. Heffron, R.J. Energy multinationals challenged by the growth of human rights. *Nat Energy* (2021). https://doi.org/10.1038/s41560-021-00906-6.
6. Heffron, R. J. 2021. *The Challenge of Energy Justice*. Springer: Heidelberg, Germany. *Forthcoming September 2021*.
7. HSBC. 2018. *Enabling a just transition to a low-carbon economy in the energy sector. Progress and lessons in Emerging Markets*. Available at: https://www.sustainablefinance.hsbc.com/carbon-transition/enabling-a-just-transition-to-a-low-carbon-economy-in-the-energy-sector (last accessed 1 September 2021).

Index[1]

A

Access, 25, 29, 34, 46, 48, 56, 91, 92, 94, 127–130, 134, 139
Affordability, 128, 136
Africa, 37n7, 91–92, 100–101, 105, 111n46, 128, 130, 132, 142n4
ASEAN Centre for Energy (ACE), 132–134, 139–140
Asia, 34, 128, 138
Australia, 3, 11, 28, 30, 33, 44–46, 54, 55, 59, 64, 66, 68, 70, 71, 88, 97, 132, 153

B

Bank, 31, 35, 127, 138, 139, 156
Business-as-usual, 6

C

Canada, 3, 33, 44, 46–47, 53, 54, 56, 57, 60, 62–64, 66, 68–71, 96, 104, 105, 117n132, 117n133, 117n134, 117–118n135, 132, 141, 153
Carbon dioxide (CO_2), 11, 13, 17, 33, 88, 111n46
Caribbean, 94–95, 103–104, 129, 139
Climate change, 6, 10, 25, 26, 35, 44, 46, 49, 50, 52, 53, 59, 60, 62–65, 88, 95–97, 101, 127, 128, 132, 134, 138, 142, 154, 156
Climate, energy, and environmental (CEE), 9–17, 22
Climate justice, 10, 14, 17
Conceptual framework, 22
Convergence, 6
Cosmopolitan justice, 24, 32–34, 63–66, 134–135
COVID-19, 5, 7, 103, 104, 142, 151
Crisis, 26, 45, 110n41, 132, 135, 142, 151–156
Critical minerals, 22, 24–36

[1] Note: Page numbers followed by 'n' refer to notes.

R. J. Heffron, *Achieving a Just Transition to a Low-Carbon Economy*, https://doi.org/10.1007/978-3-030-89460-3

D
Decommissioning, 24, 29–31, 130
Development bank, 126, 138
Distributive justice, 23, 25, 27–29, 55–58, 127–129, 131

E
Economics, 3, 11, 16, 17, 22, 26, 28, 32, 48, 49, 51, 52, 55, 57–59, 63, 65, 88–90, 94–97, 99, 101, 102, 104, 105, 109n23, 110n37, 114n87, 126, 127, 129, 136–140, 151, 152, 156
Education, 47, 51, 95, 104, 130, 136
Energy economics, 17, 102
Energy justice, 9, 10, 13, 14, 17
Energy life cycle, 10, 22
Energy system, 11, 35, 136
Environmental Impact Assessments (EIAs), 28–30, 33
Environmental Justice, 9–10, 13, 15, 97
Equality, 2, 26, 154
Equity, 26, 95, 103, 127, 132, 154
European Investment Bank (EIB), 113n77, 127, 129–132, 138, 142n4, 143n19, 144n28, 145n40, 146n63
European Union (EU), 3–5, 25, 29, 44, 47–49, 55, 67, 69, 92, 93, 101, 102, 105, 106, 131, 133, 138, 140, 142, 153
Evolution, 3, 12, 16, 126, 156
Extractive Industries Transparency Initiative (EITI), 27

F
Fairness, 2, 25, 26, 47, 128, 154
Foreign investment, 11, 121n185
Fossil fuel, 2, 3, 5, 6, 11, 15, 46, 48, 62, 63, 90, 91, 93, 96, 97, 99, 100, 102, 104, 105, 114n87, 117n123, 126, 131, 155
Future generations, 88

G
Germany, 2–4, 44, 45, 48, 53, 54, 59, 60, 63, 64, 67, 69–71, 101, 115n102, 153, 156

H
Human rights, 10, 22, 32, 33, 35, 90, 154, 155
Hydropower, 93, 102, 110n39, 114n97

I
Inclusive, 6, 12, 47, 51, 52, 127, 128, 135, 138, 141
Inequality, 11, 13, 17, 22, 26, 27, 32, 130
Infrastructure, 11, 14, 31, 47, 56, 58, 93, 94, 100, 102–104, 110n39, 113n75, 117n132, 120n170
Insurance, 127, 135
International Institute for Sustainable Development (IISD), 39n25, 42n52, 132, 141
Investment, 6, 11, 32, 36, 57, 59, 60, 89, 90, 93–95, 100, 101, 113n75, 126, 127, 136, 138, 142, 155, 156
Investors, 4, 27, 31, 32, 36, 102, 131, 155, 156
IPCC, 26
Ireland, 3, 44, 49–50, 54, 57, 61, 67, 71, 153

J
Just, 1, 2, 5, 6, 10, 11, 17, 22, 27, 30, 31, 34, 36, 44, 95, 104–106, 154, 155
JUST Framework, 6, 15, 16, 21–36, 53–70, 155
Just transition, v, vi, 1–7, 9–17, 25, 32, 44–71, 87–106, 126–142, 151–156

L
Legal cases, 31
Low-carbon, vi, 1–7, 9, 11, 12, 17, 22, 25, 26, 32, 33, 35, 36, 36n1, 44, 45, 47, 50, 52–55, 59, 61, 64, 66, 87, 92, 94–97, 100–103, 105, 120n170, 122n204, 126, 130, 132, 152, 154–156

M
Middle East, 89–90, 99
Multinational companies (MNCs), 34

N
NGO, 88, 99
North America, 95–96, 104–105, 133

O
Oil, 2, 4–6, 11, 23, 33, 90, 93, 95, 96, 99, 101, 102, 104, 110n37, 117n134, 122n204, 123n215, 135, 141
Operation, 2, 27, 28, 30, 34, 47, 56, 67, 68, 100
Organisation for Economic Co-operation and Development (OECD), 27, 39n25, 42n52, 135

P
Paris Agreement, 3, 33–36, 49, 65, 88, 102, 106, 126, 138, 139, 152, 154, 155
Pollution, 26, 97
Population, 32, 61, 62, 64, 88, 90, 92, 94, 101, 128, 131, 132, 134
Procedural justice, 24, 25, 28–29, 53–55, 126–128
Profit, 35, 96
Public policy, 5

R
Recognition justice, 24, 25, 31–32, 61–63, 131–133
Recovery, 123n209, 138, 142, 151, 156
Renewable, 5, 46, 50, 59, 88–96, 99, 100, 102, 105, 114n97, 115n108, 119n152, 122n191, 122n204, 123n215, 126, 130–132, 134, 136
Research community, 13, 154
Researchers, v, 11, 12, 15, 21, 24, 25, 44, 153
Restorative justice, 15, 24, 27–31, 53–62, 130–131
Revenue, 23, 27–29, 32, 35
Risk, 29, 51, 129–131, 138

S
Scholarship, 6, 9, 10, 12, 13, 16, 17, 22, 27, 45, 88, 153
Scotland, 3, 4, 44, 45, 51, 53, 54, 63, 68, 71, 142, 153
Security, 48, 109n23, 113n79, 135, 137, 140
Social License to Operate (SLO), 30–32

Society, 1–6, 9–13, 16, 17, 22, 24–27, 45, 49, 50, 52, 53, 55, 57, 59, 61, 63, 87, 93, 105, 131, 139, 152, 154, 156
Solar, 91, 100, 109n28, 109n32, 110n39, 111n46, 114n97
South Africa (SA), 3, 44, 51–52, 54, 58, 60, 63, 65, 66, 68–71, 91, 100, 153
South America, 94, 101, 135
Stakeholder (s), 5, 12, 28–30, 34, 45, 49, 52, 57, 58, 87, 154, 155
Subsidy (subsidies), 4–6, 57, 100, 127, 131
Sustainable, 2, 49, 51, 55–57, 59, 67–68, 71, 95, 99–101, 103, 104, 110n37, 136–141

T
Taxation, 22, 27, 28, 34
Technology, 5, 6, 15, 25, 34, 35, 94, 104, 105, 115n108, 126, 127, 131, 132, 135, 136, 141
Theoretical Framework, 22–36
Theory, 27
Transition, v, vi, 1–7, 9–17, 22, 24, 25, 32, 35, 36, 36n1, 44–71, 87–106, 126–142, 151–156
Transparency, 2, 28, 96, 117n132

U
United Kingdom (UK), 4, 11, 15, 35, 89, 97, 101, 142, 156
United Nations (UN), 24, 25, 31, 32, 44, 48–49, 136–137, 139, 141
United States (US), 3, 15, 30, 44, 126, 153
UN Sustainable Development Goals (SDGs), 24, 49, 97, 98, 110n41, 137, 141

W
Waste management, 24, 29, 30, 112n66, 130
Wind, 91, 110n39
World Bank, 35, 89, 100, 109n23, 115n102, 115n105, 130, 137

The manufacturer's authorised representative in the EU is Springer Nature Customer Service Centre GmbH, Europaplatz 3, 69115 Heidelberg, Germany. If you have any concerns regarding our products, please contact ProductSafety@springernature.com

Printed and bound by CPI Group (UK) Ltd, Croydon, CR0 4YY
30/09/2026
02225262-0003